Electromechanical Energy Conversion - I

Electromechanical Energy Conversion - I

R. C. Goel
Professor (Retired),
Electrical Engineering Department, Madan Mohan Malviya Engineering College
(now MMM Technical University) Gorakhpur-10, (UP) India
rcgoel1@yahoo.co.in

Bharti Dwivedi
Professor (Retired),
Electrical Engineering Department, Institute of Engineering &Technology
Lucknow-21, (UP) India
bharti.dwivedi@ietlucknow.ac.in

Anurag Tripathi
Professor,
Electrical Engineering Department, Institute of Engineering &Technology
Lucknow-21, (UP) India
anurag.tripathi@ietlucknow.ac.in

BSP BS Publications
An imprint of **BSP Books Pvt. Ltd.**
4-4-309/316, Giriraj Lane,
Sultan Bazar, Hyderabad - 500 095.

Electromechanical Energy Conversion - I
***by* R. C. Goel, Bharti Dwivedi and Anurag Tripathi**

***Published by*:**

BSP BS Publications
An imprint of **BSP Books Pvt. Ltd.**
4-4-309/316, Giriraj Lane,
Sultan Bazar, Hyderabad - 500 095.
Phone :040 - 23445688
e-mail: info@bspbooks.net
www.bspbooks.net

ISBN: 978-93-95038-26-3 (Hardback)

Preface

The book entitled "Electromechanical Energy Conversion–I", delves into the realm of Transformers and DC Machines. It deals with the topics of Electrical Machines, extracted from the syllabi followed in almost all the Universities of India. The idea of writing this book has been felt by the experienced authors seeing certain paucities in the books available in the market. The authors felt that there is a need of explaining the fundamentals to the students in a more lucid manner than what is there in the existing books. The material in existing books is either too concise or too elaborate for readers' understanding. The long-standing experience of the authors in teaching this subject has helped them in envisioning the issues the students find difficult to understand using the available books. Electrical Machines is a course which needs clear understanding of its fundamentals, which eventually involves basic, yet complex, underlying concepts. Once the student understands the basic concepts, the subject becomes interesting to the student. This results in student's enhanced problem-solving capability and confidence in the subject. Authors have attempted to maintain brevity while ensuring clarity at the same time.

This book contains five chapters covering concepts of Electromechanical Energy Conversion-I, Single Phase Transformers, Three Phase Transformers, D.C. Generators and D.C. Motors. As an illustration take the example of chapter dealing with Single Phase Transformer. Effort has been made to give an immediate impact and understanding about the transformer. A very simple and basic diagram of single-phase transformer is given distinctly making every aspect clear. By reading the Introduction, the fundamentals of the transformers would be clear. One can understand its principles and construction, EMF equation has been developed making it clear that the EMF induced wave lags behind the flux wave by 90 degrees. This is a very important concept and has wide intricate implications. Another very important and useful example is that of power flow analysis for a transformer; which has been explained in a very simple manner in this book. Similarly, the complex concept involved in analysis of 3-phase transformers has been dealt with in a very simple to understand manner. 3-phase transformer Phasor Groups have been discussed in an extremely lucid manner which one may not find in most of the similar books. The Phenomena of Excitation and Harmonics in transformers have also been considered adequately.

Further, starting from the constructional details of DC machines, and going up to their operational and performance details, the book offers an easy path to develop the understanding of the machines.

Thus, by going through the book one will find that it really includes outstanding pedagogical features with suitable illustrative examples and problems.

- Authors

Contents

CHAPTER 3

Three Phase Transformer

CHAPTER 4

DC Generators

CHAPTER 5

Dc Motors

CHAPTER 1
ELECTROMECHANICAL ENERGY CONVERSION

1.1 Electromechanical Energy Conversion Devices

An electromechanical energy conversion device is the one which converts mechanical energy into electrical energy or vice-versa.

Generator - Converts mechanical energy into electrical energy

Motor - Converts electrical energy into mechanical energy

The block diagram in Fig.1.1 represents an electromechanical energy conversion system. The electrical and the mechanical systems are interconnected through a magnetic field coupling. Such a system can perform bi-lateral conversion i.e., conversion of mechanical energy into electrical energy or vice versa.

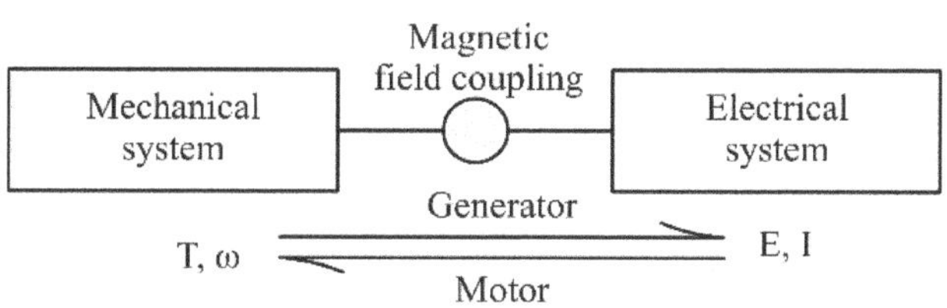

Figure 1.1 Energy Conversion

1.1.1 Generator Action

An electromechanical energy conversion device converting mechanical energy into electrical energy is shown in Fig.1.2.

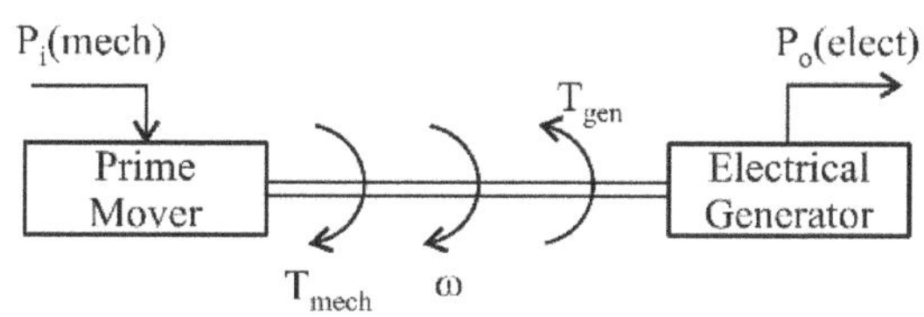

Figure 1.2 Generator Action

The mechanical energy input $P_{i(mech)}$ is supplied to a prime mover. The prime mover is mechanically coupled with the electrical generator to transmit mechanical energy to the generator. A mechanical torque T_{mech} is developed by the prime mover which rotates the generator at a speed ω rad/sec. Electrical energy output $P_{o(elect)}$ is developed by the generator. The generator torque T_{gen} balances the mechanical torque T_{mech}. The energy balance equation is given by

$P_{i(mech)}$	=	$P_{o(elect)}$	+	P_{field}	+	P_{heat}	
(Input)		(output)		(Increase in field energy)		(Cu, iron, mechanical losses converted into heat)	.....(1.1)

1.1.2 Motor Action

In this case, electrical power drives a motor and the motor drives a mechanical load as shown in Fig.1.3.

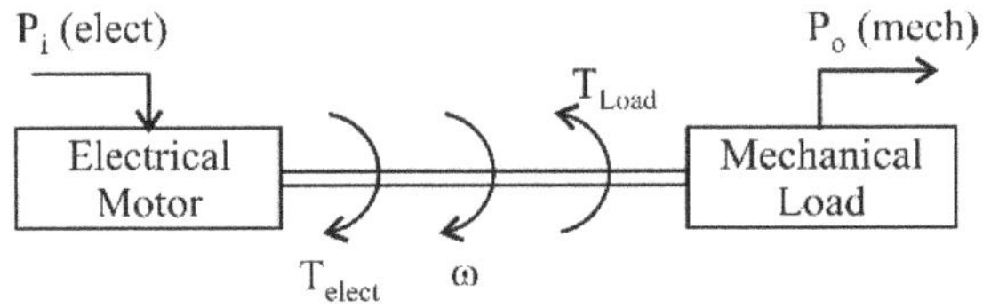

Figure 1.3 Motor Action

The energy balance equation is given by

$P_{i(elect)}$	=	$P_{o(mech)}$	+	P_{field}	+	P_{heat}	
(Input)		(output)		(Increase in field energy)		(Cu, iron, mechanical losses converted into heat)	.....(1.2)

1.2 Principle of Electromechanical Energy Conversion

It is clear from above discussion that a bilateral flow of energy is possible between an Electrical system and a Mechanical system through a common medium called Magnetic field coupling.

In all rotating machines there is an important role of electromagnetic energy conversion. A rate of change of flux is essentially needed to produce dynamically induced emf. In general,

$$\text{Induced emf} = Blv$$

where 'B' is flux density in wb/m^2, 'l' is the length of conductor in metre placed in magnetic field and 'v' is the velocity of the movement of that conductor in m/sec.

Lorentz Law: If a particle, having charge of 'q' coulombs is moved in a pure electric field E volt/m, then the force experienced is given by

$$F = qE \text{ (in pure electric field)} \quad .....(1.3)$$

Now, if the same charged particle is moved in a pure magnetic field having flux density of 'B' wb/m^2, with a velocity of 'v' m/sec, then the force experienced is given by

$$F = gv \times B \text{ (in pure magnetic field)} \quad .....(1.4)$$

The Lorentz law states that the force experienced by a particle charge in the presence of an electric and a magnetic field together is given by

$$F = q\,(E + v \times B) \qquad \text{.....(1.5)}$$

If the number of charged particles moving is large, then the force is given by

$$F_V = \rho\,(E + v \times B)$$

where F_V is force density (force/velocity is N/m^3) and 'ρ' is charge density (coloumb/m^3)

Also, $$J = \rho v \qquad \text{.....(1.6)}$$

where J is the current density (Amp/m^2)

Thus, in a pure magnetic field, the force is given by

$$F_V = J \times B \qquad \text{.....(1.7)}$$

1.3 Electromechanical Force

An electromechanical force is commonly seen to appear in following cases:

1. In alignment of magnetic fields:
   (a) When a magnetic material is placed under the influence of a magnetic field Fig.1.4 (a) or
   (b) When a weak magnet is placed under the influence of a stronger magnetic field Fig.1.4(b)

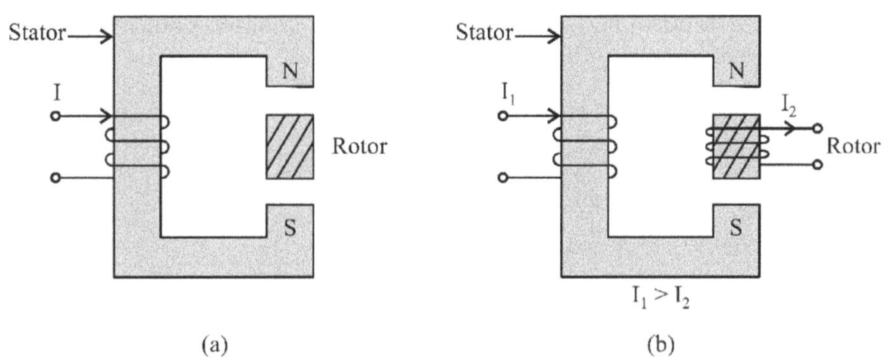

Figure 1.4 Rotor Alignment

2. When a current carrying conductor is placed in a magnetic field (motor action) or when a conductor is moved in a magnetic field (generator action). The above actions are explained through Flemings Left Hand Rule and Fleming's Right Hand Rule respectively.

1.4 Energy Flow in Electromechanical Systems

The flow of energy can be understood by following:

$$\begin{bmatrix}\text{Electrical}\\ \text{Energy Input}\end{bmatrix} \equiv \begin{bmatrix}\text{Mechanical}\\ \text{Energy}\\ \text{Output}\end{bmatrix} + \begin{bmatrix}\text{Increase}\\ \text{in Field}\\ \text{Energy}\end{bmatrix} + \begin{bmatrix}\text{Energy lost}\\ \text{in the form of}\\ \text{heat } (I^2R)\end{bmatrix} \quad \text{.....(1)}$$

The above relationship can be rearranged as

$$\begin{bmatrix}\text{Electrical}\\ \text{Energy}\\ \text{Input}\\ \text{minus } (I^2R)\\ \text{Losses}\end{bmatrix} \equiv \begin{bmatrix}\text{Mechanical}\\ \text{Energy}\\ \text{Output}\\ \text{Plus}\\ \text{Friction and}\\ \text{Windage losses}\end{bmatrix} + \begin{bmatrix}\text{Increase}\\ \text{in}\\ \text{Field}\\ \text{Energy}\end{bmatrix} \quad \text{.....(2)}$$

The above can also be represented by following diagram.

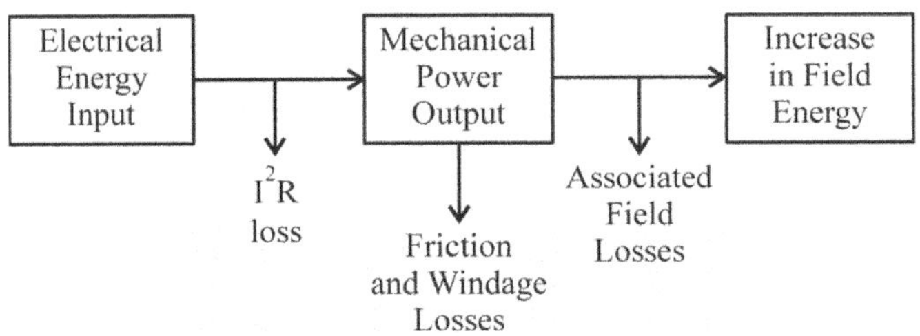

.....(3)

Based on the flow diagram (3), the differential energy balance equation can be written as follows.

$$dW_e = dW_m + dW_f \quad \text{.....(1.8)}$$

Here dW_e is the differential electrical energy supplied in time 'dt'

dW_m is the differential mechanical energy output in time 'dt'

dW_f is the differential field energy output in time 'dt'

Under steady state condition, the energy balance equation becomes

$$W_e = W_m + W_f \quad \text{.....(1.9)}$$

1.5 Energy Balance in Electromechanical Systems

This topic would be studied under following subcategories.

Singly Excited System (Single Source of Electrical Power)

A	-	Without a rotor
B	-	With a rotor
B.1	-	static energization
B.2	-	Dynamic energization
B.2.1	-	Slow rotor movement

B.2.2 - Fast (instantaneous rotor movement)

B.2.3 - Transient rotor movement

Doubly Excited System (Two Sources of Electrical Power)

1.5.1 Singly Excited System (without a Rotor)

The electrical energy supplied by a source gets either dissipated, converted into other form, or stored in magnetic or in static (like in battery, capacitor etc.) forms.

The current flowing in circuit shown in Fig.1.5 results in dissipation of a part of the energy in the resistance of coil, while the rest of the energy gets stored in magnetic field.

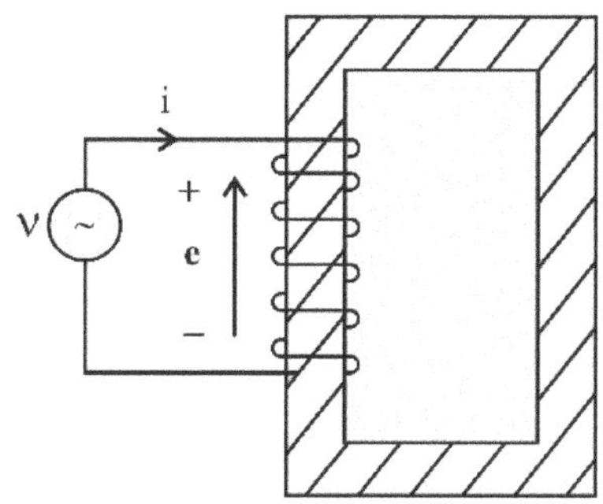

Figure 1.5 Singly Excited System without Rotor

The KVL gives $v = e + iR$ .....(1.10)

where R is the coil resistance

The power equation can be given as

$$vi = ei + Ri^2 \quad .....(1.11)$$

The energy for duration 'dt' is given as

$$vidt = eidt + Ri^2\, dt \quad .....(1.12)$$

Here, vidt is the input energy, eidt is the energy stored in magnetic field, and Ri^2 dt is the energy lost in the form of heat.

Thus, the energy stored $dw_f = eidt$

But, $e = \frac{d\psi}{dt}$, where $\psi = N\phi$ .....(1.13)

$$\therefore \quad dw_f = \frac{d\psi}{dt} . i . dt = i\, d\psi \quad .....(1.14)$$

Or, $$\text{energy stored} = \int dw_f = \int i d\psi = \int N i d\phi \quad .....(1.15)$$

At the same time, since

$$E = L\frac{di}{dt} \quad .....(1.16)$$

Eq. (1.13) can be written as

$$dW_f = L\frac{di}{dt}.idt = Lidi$$

or, energy stored = $\int dW_f = \int Lidi$ .....(1.17)

1.5.1.1 Concept of Energy and Coenergy

Eq. (1.15) can be graphically represented as follows.

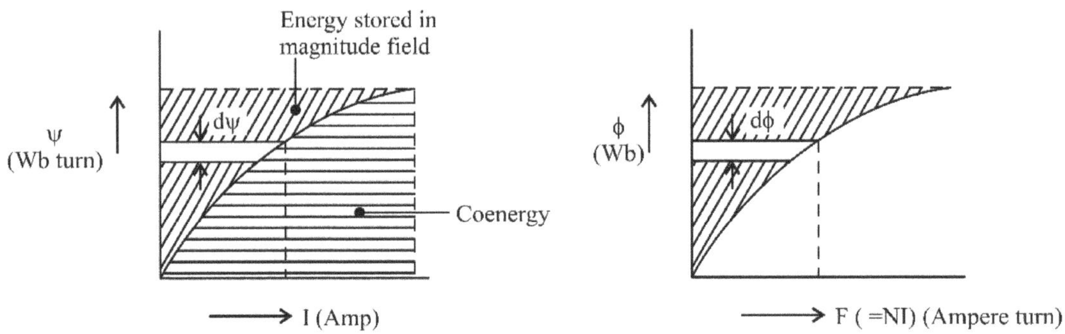

Figure 1.6 Energy - Coenergy

The variations of 'ψ' and 'ϕ' are plotted against 'I' and 'F' respectively. The area above the curve represents the amount of energy stored in magnetic field, while that under the curve is called the 'Coenergy".

Energy and Co-energy for Linear B-H curve

For Linear systems LI = Nϕ

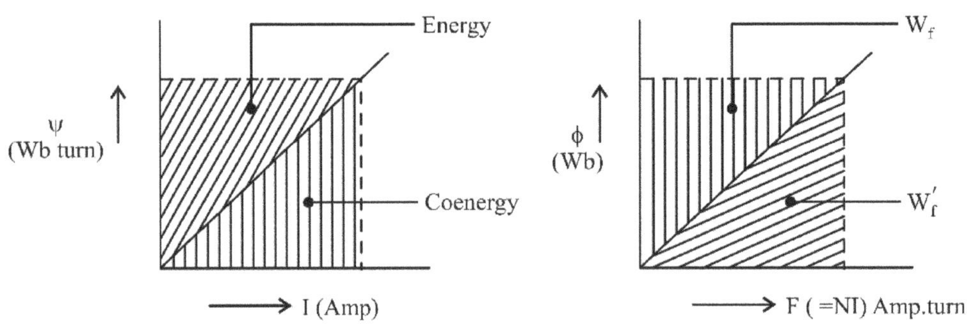

Figure 1.7 Energy – Coenergy (Linear)

Let the Energy be represented by W_f and the Coenergy by W_f'.

1.5.2 Singly Excited System (with Rotor)

The ferromagnetic rotor is placed at position A (Fig. 1.8). The magnetic material is energized by allowing a current to flow through the winding as a result of it, the rotor moves by angle θ_m and acquired position B. The torque acting on the Motor to make this movement is called saliency (Reluctance) torque. It is the torque exerted on rotor to bring the rotor to a position where the reluctance of magnetic path is minimum.

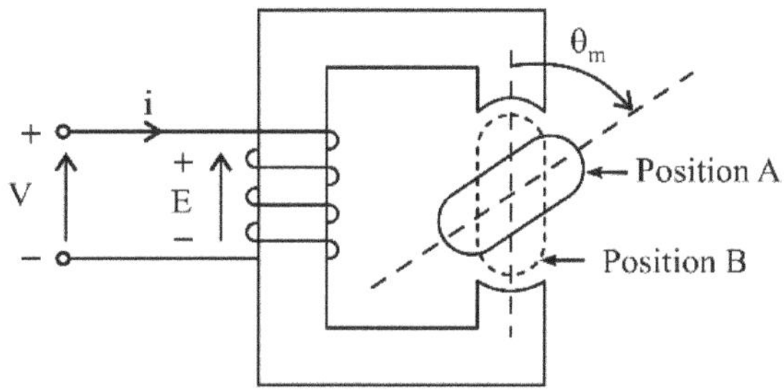

Figure 1.8 Singly excited system with rotor

The assumptions made in the analysis to follow are as given below.

1. ψ - i relationship is linear

   That is $\psi = L(\theta)\, i$
2. No iron loss takes place
3. Leakage flux is zero
4. Electrostatic field is not present
5. The core material of stator and rotor has $\mu_r = \infty$. That is, the reluctance offered by magnetic path is zero. In other words, the net reluctance in the magnetic path is caused by the air gap only.
6. The flux crossing the air gap is uniform, undistorted and radial with no fringing effect.
7. The air gap is very thin so that constant flux density is maintained in the air gap.

If there is no reluctance in the magnetic core, then the total available mmf will be utilized equally in creating air gap flux in the two parts of airgap.

The voltage equation is given by

$$V = Ri + {}^{d\psi}\!/_{dt} \qquad \text{.....(1.18)}$$

The energy is represented by

$$W_e = \underbrace{\int_0^t vidt}_{\substack{\text{Energy}\\ \text{Input}}} = \underbrace{\int_0^t i^2Rdt}_{\substack{\text{Energy}\\ \text{lost}(=W_l)}} + \underbrace{\int_0^t \frac{id\psi}{dt}dt}_{\substack{\text{Useful}\\ \text{Energy}(=W_u)}} \qquad \text{.....(1.19)}$$

$W_e = W_l + W_u$

$W_u = W_f + W_m$

where W_f = energy stored in magnetic field

W_m = energy converted to mechanical energy output

1.6 Static Energization

Let us analyze the situation where the rotor occupies two static positions, position A and position B, as shown in Fig.1.8. The $\psi - I$ characteristics for these two positions are drawn in Fig.1.9.

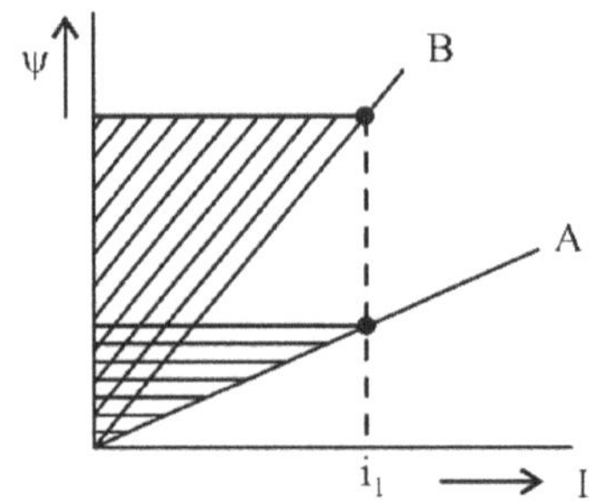

Figure 1.9 Energy Stored for Fig.1.8

The energy stored in magnetic field when the rotor occupies positions A and B are shown by the shaded regions in Fig.1.9.

As the rotor is stationary, $W_m = 0$; No mechanical energy exchange is involved. Therefore

$$\int_0^{\psi} i d\psi = W_f \text{ ; energy stored}$$

or
$$W_f = \int_0^{\psi} i d\psi = \int_0^{\psi} \frac{\psi}{L} d\psi \text{ ; as } \psi = N\phi = Li$$

$$= \frac{\psi^2}{2L} = \frac{L^2 i^2}{2L} = \frac{1}{2} Li^2$$

Thus, the energy stored is given by $\frac{1}{2}Li^2$. Further the slope of ψ - I characteristic is given by

$$\frac{\psi}{i} = \frac{N\phi}{i} = L \qquad \text{.....(1.20)}$$

Figure 1.9 shows that $L_A < L_B$.

It means that the inductance and hence the energy stored, is maximum when the rotor is aligned with the static magnetic field (Position B). The value of inductance and energy stored keep reducing as the rotor occupies positions away (e.g. Position A) from the axis of static magnetic field, for the same amount of current in the magnetic coil.

1.7 Dynamic Energization

A situation is considered where the rotor moves from position A to position B in Fig.1.10.

This movement can take place either slowly or instantaneously.

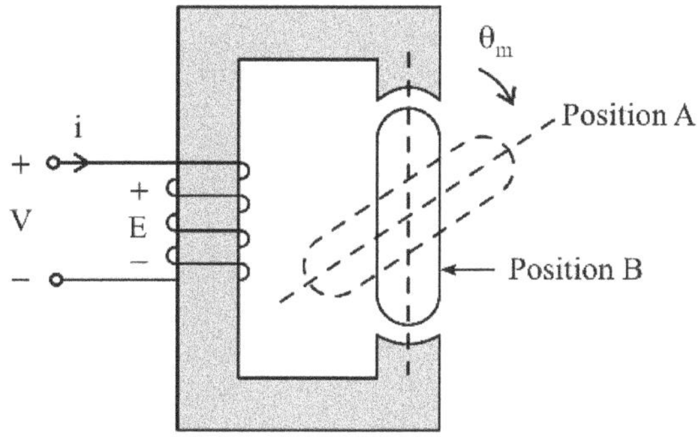

Figure 1.10 Dynamic Energization

1.7.1 Slow Movement of Rotor

The rotor slowly moves from position A to B. This movement is so slow that the rate of change of flux experienced by it is small. Therefore, no emf gets induced. Hence there is no change in current. The change in energy magnetic field (ΔW_f) is the difference between the energy stored at rotor position A (W_{fA}) and that at rotor position B (W_{fB}). As shown in Fig.1.11.

$$\Delta W_f = W_{fB} - W_{fA}$$

$$= \Delta Oef - \Delta Odc$$

$$= \frac{1}{2}[oe \times ef) - (od \times ef)]$$

$$= \frac{1}{2} ef (oe - od) = \frac{1}{2} ef \times de \qquad .....(1.21)$$

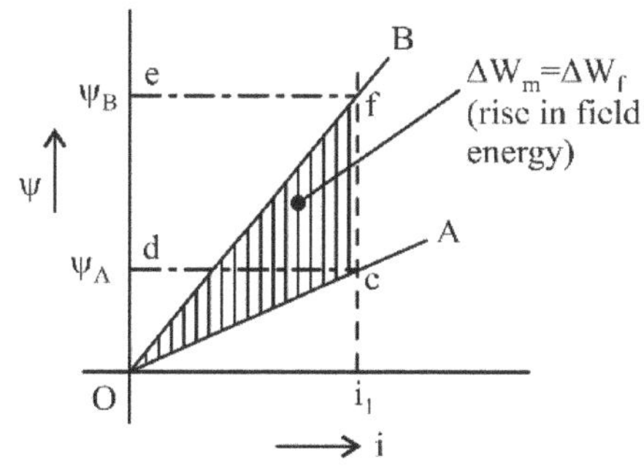

Figure 1.11 Energy Stored – Slow motion

$$\Delta W_e = \int_{\psi A}^{\psi B} i_1 d\psi = i_1(\psi_B - \psi_A) = ef \times de \qquad ....(1.22)$$

This change in electrical change input (ΔW_e) leads to rise in magnetic field energy (ΔW_f) and in mechanical energy (ΔW_m) to shift the rotor position.

Thus, the mechanical energy ΔW_m spent in slowly moving the rotor from position A to B results in rise of ΔW_f.

Expression for Electromechanical Torque T_m

$$T_m = \lim_{\theta_m \to 0} \left\{ \frac{\Delta W_f}{\Delta \theta_m} \right\} \Bigg|_{\text{constnat } i} \qquad \text{.....(1.23)}$$

$$= \frac{\partial W_f}{\partial \theta_m} \Bigg|_{\text{constant } i}$$

$$= \frac{\partial}{\partial \theta_m} (W_f) \Bigg|_{\text{constant } i}$$

We know that $W_f = \frac{1}{2} L i^2$

Also, the system is assumed to be linear,

Hence $L \neq f(i)$

$$\therefore \qquad T_m = \frac{i^2}{2} \frac{dL}{d\theta_m} \qquad \text{.....(1.24)}$$

1.7.2 Fast Movement of Rotor

The rotor instantaneously moves from position A to position B as shown in Fig.1.12. The flux linkage however cannot change instantaneously and remains the same as ψ_1 during rotor movement. In Fig.1.12 it is depicted that the initial operating point is at c. With the sudden movement of rotor, it gets shifted to c′ at the same flux linkage ψ_1. After the rotor reaches at position B, the operating point shifts from c′ to c. Current i_2 is drawn to develop ψ_1 flux linkage and the operating point settles down at c for rotor position B. The expression of electrical energy W_e is given by.

$$W_e = \int_{\psi_1}^{\psi_2} i \, d\psi$$

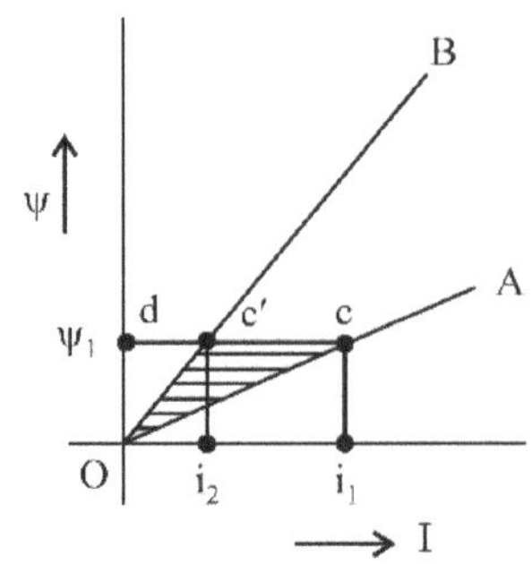

Figure 1.12 Energy Stored – Fast movement

Since due to fast rotor movement, the flux linkage does not change, therefore $\psi_c = \psi_{c'} = \psi_1$ is constant.

So, $W_m = -W_f$ (decrease in stored magnetic field energy)

It shows that no electric energy is taken during fast movement of the rotor. The stored magnetic energy only gets consumed in such a rotor movement. It is shown by occ' in Fig.1.12.

Expression for torque T_m is given by

$$T_m = \lim_{\theta_m \to 0} \left\{ \frac{-\Delta W_f}{\Delta \theta_m} \right\} \Bigg|_{\text{constant } \psi} \qquad \text{.....(1.25)}$$

$$= \frac{-\partial W_f}{\partial \theta_m} \Bigg|_{\text{constant } \psi} = \frac{-\partial}{\partial \theta_m} \left(\frac{\psi^2}{2L} \right) \Bigg|_{\text{constant } \psi}$$

$$\therefore \qquad T_m = \frac{-\psi^2}{2} \frac{1}{L^2} \frac{\partial L}{\partial \theta_m}$$

$$= \frac{i^2}{2} \frac{\partial L}{\partial \theta_m}$$

For linear system

$$T_m = \frac{i^2}{2} \frac{dL}{d\theta_m} \qquad \text{.....(1.26)}$$

1.7.3 Transient Rotor Movement

Between the two extremes, actually when the rotor moves from A to B, the speed is neither very slow nor very fast. The mechanical energy used by the rotor is given by area oce (shaded). The current gets reduced from i_2 to i_1 as the rotor gets aligned to magnetic field and the flux linkage rises from ψ_1 to ψ_2. In Fig.1.13 the graphical relationship gives the following change in electrical energy

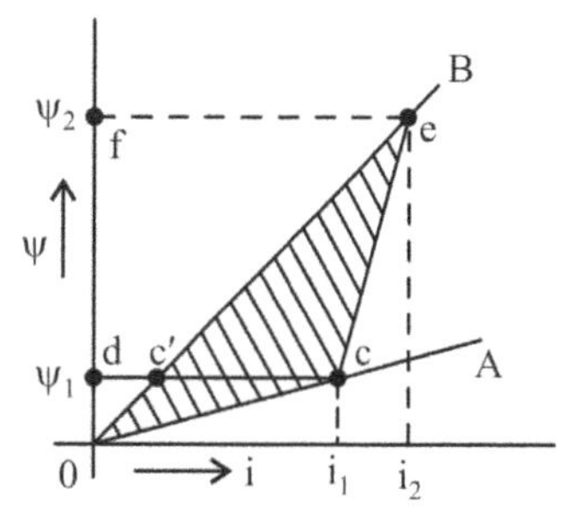

Fig.1.13 Transient Rotor Movement

ΔW_e = fecd (input energy)

change in field energy

ΔW_f = Oef – Ocd (stored energy).

Hence change in mechanical energy = $\Delta W_e - \Delta W_f$ (output)

or $\Delta W_m = \Delta W_e - \Delta W_f$

= fecd – Oef + Ocd = Ocef – Oef

= Oce (shaded portion)

1.8 Doubly Excited System

In doubly excited system both the stator and the rotor are excited as shown in Fig.1.14. The rotor and stator have saliency. The assumptions made in carrying out this analysis are same as given earlier.

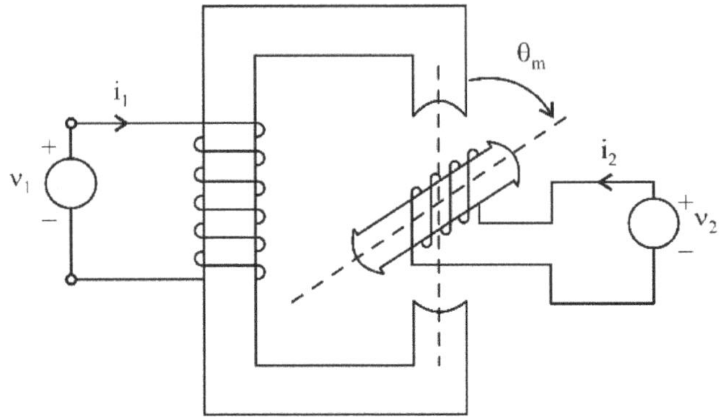

Figure 1.14 Doubly Excited System

Equation set (1) – Flux linkages

$$\psi_1 = L_1 i_1 + Mi_2$$

$$\psi_2 = L_2 i_2 + Mi_1$$

Equation set (2) voltages

$$v_1 = R_1 i_1 + \frac{d\psi_1}{dt}$$

$$v_2 = R_2 i_2 + \frac{d\psi_2}{dt}$$

Equation set (3) – Combining equation sets (1) and (2)

$$v_1 = R_1 i_1 + L_1 \frac{di_1}{dt} + i_1 \frac{dL_1}{dt} + i_2 \frac{dM}{dt} + M \frac{di_2}{dt}$$

$$v_2 = R_2 i_2 + L_2 \frac{di_2}{dt} + i_2 \frac{dL_2}{dt} + i_1 \frac{dM}{dt} + M \frac{di_1}{dt}$$

Equation set (4) – Power equations (Multiplying (3) with i)

$$v_1 i_1 + R_1 i_1^2 + L_1 i_1 \frac{di_1}{dt} + i_1^2 \frac{dL_1}{dt} + i_1 i_2 \frac{dM}{dt} + Mi_1 \frac{di_2}{dt}$$

$$v_2 i_2 + R_2 i_2^2 + L_2 i_2 \frac{di_2}{dt} + i_2^2 \frac{dL_2}{dt} + i_1 i_2 \frac{dM}{dt} + Mi_2 \frac{di_1}{dt}$$

Equation set (5) – Energy equations (integrating and adding equation sets in (4))

$$\int_0^t (v_1 i_1 + v_2 i_2) dt = \int_0^t \underbrace{\left(R_1 i_1^2 + R_2 i_2^2\right)}_{[A]} dt$$

$$+\underbrace{\int_0^t \left(L_1 i_1 \frac{di_1}{dt} + L_2 i_2 \frac{di_2}{dt} + M i_1 \frac{di_2}{dt} + M i_2 \frac{di_1}{dt} \right) dt}_{[B]}$$

$$+\underbrace{\int_0^t \left(i_1^2 \frac{dL_1}{dt} + i_2^2 \frac{dL_2}{dt} + 2 i_1 i_2 \frac{dM}{dt} \right) dt}_{[C]}$$

The above equation has three components and they express the energies. [A] gives energy lost as copper loss. [B] is stored energy in magnetic field and [C] is energy converted into mechanical form.

[A] corresponds to energy lost in copper windings on stator and rotor. This is given by

$$\int_0^t \left(R_1 i_1^2 + R_2 i_2^2\right) dt$$

[B] corresponds to energy stored in magnetic field W_f and is given by

$$W_f = \int_0^t \left(L_1 i_1 \frac{di_1}{dt} + L_2 i_2 \frac{di_2}{dt} + M i_1 \frac{di_2}{dt} + M i_2 \frac{di_1}{dt} \right) dt$$

[C] corresponds to mechanical energy developed

The term $\left(i_1^2 \frac{dL_1}{dt} + i_2^2 \frac{dL_2}{dt} + 2 i_1 i_2 \frac{dM}{dt} \right)$ is zero because the system considered is linear

Part [B] of the expression is used to obtain expression for torque.

The expression for Electromagnetic torque (T_e)

$$T_e = \frac{dW_f}{dt}$$

$$= \frac{d}{dt}\left[\frac{1}{2} L_1 i_1^2 + \frac{1}{2} L_2 i_2^2 + 2 M i_1 i_2 \right]$$

$$i_1 = \begin{cases} \frac{1}{2} L_1 \frac{di_1^2}{dt} + \frac{1}{2} L_2 \frac{di_2^2}{dt} + 2 M i_1 \frac{di_2}{dt} + 2 M i_2 \frac{di_1}{dt} \\ +2 i_1 i_2 \frac{dM}{dt} + \frac{1}{2} i_1^2 \frac{dL_1}{dt} + \frac{1}{2} i_2^2 \frac{dL_2}{dt} \end{cases}$$

$$di_2^2 = \begin{Bmatrix} L_1 i_1 \dfrac{di_1}{dt} + L_2 i_2 \dfrac{di_2^1}{dt} + 2Mi_1 \dfrac{di_2}{dt} + 2Mi_2 \dfrac{di_1}{dt} \\ +2i_1 i_2 \dfrac{dM}{dt} + \dfrac{1}{2} i_1^2 \dfrac{dL_1}{dt} + \dfrac{1}{2} i_2^2 \dfrac{dL_2}{dt} \end{Bmatrix}$$

The expression for electromechanical torque (T_m)

$$T_m = \frac{dW_f}{d\theta_m}$$

$$= \left[\frac{1}{2} i_1^2 \frac{dL_1}{d\theta_m} + \frac{1}{2} i_2^2 \frac{dL_2}{d\theta_m} + \left(2i_1 i_2 \frac{dM}{d\theta_m} \right) \right]$$

In above equation the term $\left(\frac{1}{2} i_1^2 \frac{dL_1}{d\theta_m} + \frac{1}{2} i_2^2 \frac{dL_2}{d\theta_m} \right)$ represents the reluctance (or saliency) torque. This term becomes zero in machines having uniform air gap.

The second term $\left(2i_1 i_2 \frac{dM}{d\theta_m} \right)$ represents the co-alignment torque.

$T_m = \frac{1}{2} i^2 \frac{dL}{d\theta_m}$; if either i_1 or i_2 becomes zero. T_m thus becomes same as that of a singly excited system.

1.9 Physical Concept of Force and Torque Production

1.9.1 DC Machines

The electromagnetic field poles are present on the stator, while the armature winding is placed on the rotor. Fig.1.15 shows the Generator and the Motor action of the machine.

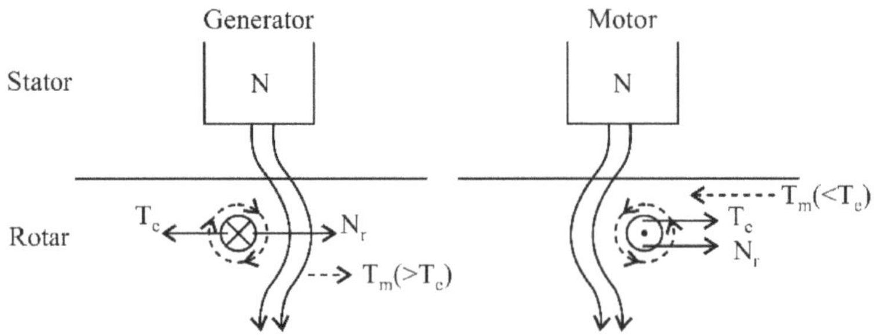

Figure 1.15 DC Machines

In case of Generator the rotor is rotated at N_r rpm. Flemings Right Hand (FRH) rule determines the current direction in armature conductor. The main field flux gets deviated due to magnetic field surrounding the armature conductor.

The magnetic lines of forces are elastic in nature. It can be assumed that they try to straighten themselves, thus developing an electromagnetic torque T_e to work opposite to the direction of rotation. In generators, the mechanical torque T_m applied externally on the rotor. T_m is greater than the electromagnetic torque T_e developed.

T_m and T_e work opposite to each other. Hence in case of generators, the rotation N_r, resulted due to T_m is in opposite direction to T_e.

In case of Motor, the armature conductor carries current in the direction shown. Applying Fleming's Left Hand (FLH) rule, the direction of rotation obtained is in the same direction as that of T_e. As the load on the motor is mechanical, the load torque T_m will work opposite to T_e. However, $T_e > T_m$ makes the motor run at N_r in the same direction as that of T_e.

1.9.2 Synchronous Machines

The arrangement of the field and the armature windings in synchronous machines are opposite to that of DC machines, as shown in Fig.1.16.

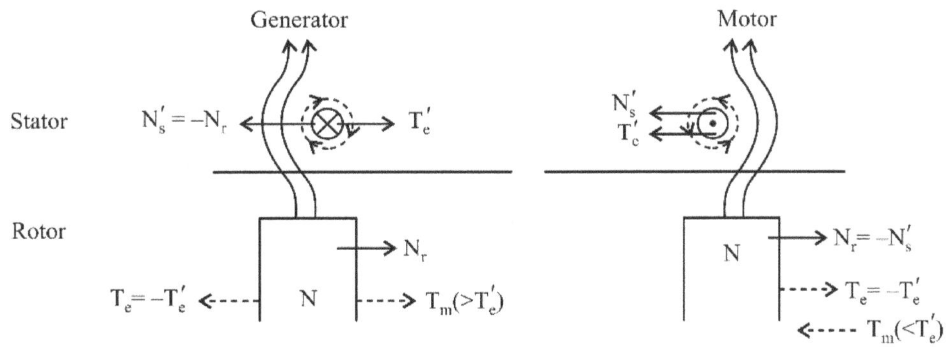

Figure 1.16 Synchronous Machines

In case of Generator, the rotor is rotated at N_r. In other words, the armature conductor moves at a fictitious speed of $N'_S(=-N_r)$. The FRH rule determines the direction of current in armature conductor. The surrounding magnetic field of which deviates the main field flux as shown. These lines of forces develop a torque T'_e on armature conductors. In other words, an equal and opposite torque T_e acts the rotor. The rotor rotates at N_r because the mechanical input torque T_m is greater than T_e.

Similarly, in case of motor, the armature current carries a current as shown. The FLH rule gives the direction of movement of armature conductor as N'_S (fictitious). The magnetic field surrounding the armature conductor, deviates the main field flux such that a fictitious electromagnetic torque gets developed as T'_e. As the armature conductor cannot move, the rotor moves at N_r $(=-N'_S)$ due to the torque $T_e(=-T'_e)$ acting indirectly on it. The mechanical torque T_m opposes the electromechanical torque T_e in both the cases.

1.9.3 Induction Machines

The stator and rotor carry three phase ac distributed field and armature windings respectively. Fictitious poles are created in both the stator and rotor. The poles created due to stator winding rotate at N_s rpm (synchronous speed), while the rotor poles rotate at N_r rpm.

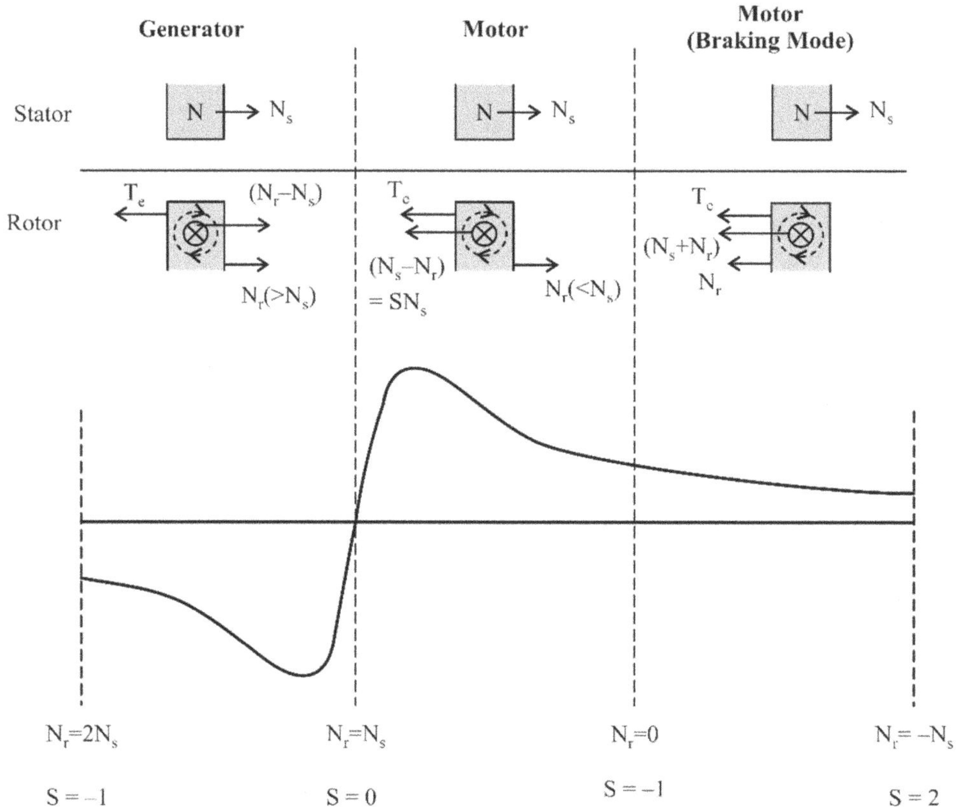

Figure 1.17 Induction Machine

Fig.1.17 explains the operation of induction machine in all its three modes; i.e., Generating, Motoring and Braking modes. The corresponding slip-torque characteristic is also drawn for the three modes of operation of induction machine.

1.9.4 Summary

The above discussed can be summarized as follows.

1. The direction of current in the armature conductor decides the direction of its surrounding magnetic field. This causes deviation in the main field flux. The main field flux is stronger, and due to the basic elastic nature of flux, it tries to straighten itself. Moreover, the weaker field, surrounding the armature conductor, tries to align itself with the main field flux. In this process, the armature conductor experiences force resulting in development of T_e.

2. In case of Generators, the electromechanical (input) torque (T_m) is greater than the electromagnetic (load) torque (T_e). As the rotor rotates due to the T_m (input), the direction of rotation of rotor is the same as that of T_m. Hence T_e and N_r act in opposite directions.
3. In case of Motors, the electromagnetic (input) torque (T_e) is greater than the Electromechanical (load) torque (T_m). The rotor rotates due to T_e, hence T_e and N_r act in the same direction.

REVIEW QUESTIONS

1. What is meant by Electromechanical Energy Conversion?
2. Describe the concept of Energy and Co-energy.
3. Derive the expression for torque developed in doubly excited system. From this expression derive the expression for torque in singly excited system.
4. With the help of diagrams, explain the effect of saturation on stored magnetic energy.
5. How does an Induction machine change its behavior when the rotor speed is varied from negative synchronous speed to positive synchronous speed?

PROBLEMS

1. The following are the inductances in a doubly excited system.

   $L_1 = 10 + 5 \cos 2\theta$ henry

   $L_2 = 9 + 4 \cos 2\theta$ henry

   $M = 15 \cos \theta$ henry

   The currents in the two coils are

   $I_1 = 0.5$ Amp

   $I_2 = 0.7$ Amp

Calculate

(i) Torque as a function of θ and its value for $\theta = -45°$

(ii) Expression for energy stored as a function of θ.

Sol: (i) The expression of torque is as given below

$$T = \frac{1}{2} i_1^2 \frac{dL_1}{d\theta} + \frac{1}{2} i_2^2 \frac{dL_2}{d\theta} + i_1 i_2 \frac{dM}{d\theta}$$

$$= \frac{1}{2}(0.5)^2 \frac{d}{d\theta}(10 + 5\cos 2\theta) + \frac{1}{2}(0.7)^2 \frac{d}{d\theta}(9 + 4\cos 2\theta) + (0.5)(0.7)\frac{d}{d\theta}(15\cos\theta)$$

$$T = \frac{1}{2}(0.5)^2(-10\sin 2\theta) + \frac{1}{2}(0.7)^2.(-8\sin 2\theta) + (0.5)(0.7)(-15\sin\theta)$$

$$= -3.21 \sin(90) - 5.25 \sin(45)$$

or $T = -6.91$ nm

(ii) Energy stored in field as a function of θ

$$Wf_\theta = \frac{1}{2}L_1 i_1^2 + \frac{1}{2}L_2 i_2^2 + Mi_1 i_2$$

$$= \frac{1}{2}(10 + 5\cos 2\theta)(0.5)^2 + \frac{1}{2}(9 + 4\cos 2\theta)(0.7)^2$$

$$+ (15\cos\theta)(0.5)(0.7)$$

$$= 3.455 + 1.605\cos 2\theta + 5.25\cos\theta$$

CHAPTER 2

SINGLE PHASE TRANSFORMER

2.1 Introduction

Transformer is a static device to step up or step down an alternating voltage by principle of electro-magnetic induction. In its simplest form the transformer consists of a laminated silicon steel core having two different windings wound as shown in Fig.2.1.

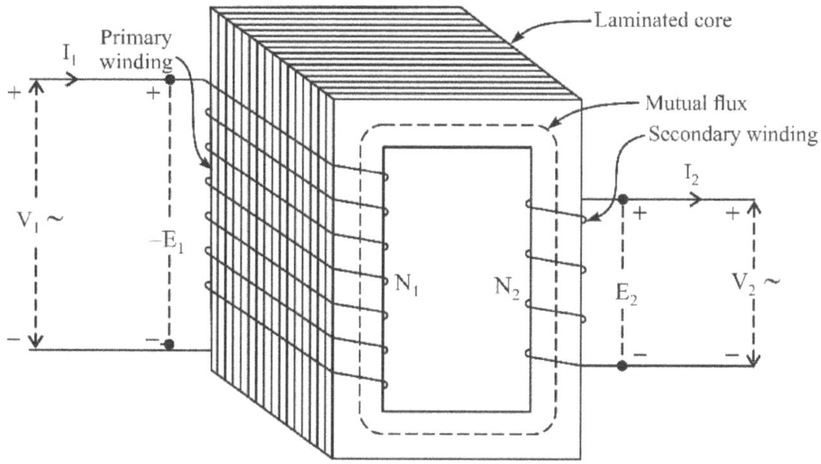

Figure 2.1 Simple 1-phase Two Winding Transformer

The winding to which alternating voltage is applied, is termed primary winding and from which output voltage taken is called secondary winding. In practice the windings are termed as High Voltage (HV) and Low Voltage (LV) windings. Transformers have wide range of application in transmission and distribution of electrical energy. For same amount of power transfer, the line losses get reduced by increasing the voltage for transfer of energy.

2.2 Principle of Operation

When an alternating voltage V_1 is applied to the primary winding of a transformer Fig.2.1, keeping secondary winding open, a small alternating current I_o flow in it. This alternating current produces an alternating flux in the laminated core which links with both the windings. This alternating flux linking the primary winding induces an alternating EMF E_1 by self-induction opposing the supply voltage V_1. The same alternating flux, also linking with the secondary winding, induces an EMF E_2 by mutual induction in it. Since secondary winding is open circuited, the secondary terminal voltage V_2 remains equal to E_2. For an ideal transformer the voltage transformation ratio k would be:

$$k = \frac{V_2}{V_1} = \frac{E_2}{E_1} = \frac{N_2}{N_1}$$

where N_1 and N_2 are number of turns in primary and secondary windings. It is assumed that in an ideal transformer, there are no losses in electric circuits and magnetic circuit.

2.3 Constructional Features

A single phase transformer has a silicon steel laminated core with primary and secondary windings as shown in Fig.2.1. There are two types of core construction.

1. Core type
2. Shell type

2.4 Core Type Construction

This type of construction has a rectangular core as shown in Fig.2.2. The vertical limbs are called core upon which windings are placed and the horizontal part is called yoke and its function is to complete flux path.

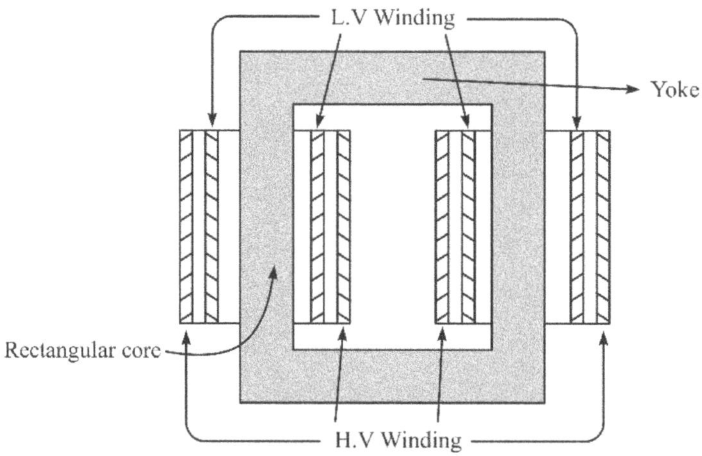

Figure 2.2 Single Phase Core Type Transformer

The primary and secondary windings are divided into two equal parts. Half of each Low voltage (LV) windings are first placed near to the core and then over the LV windings, the HV windings are placed. This construction ensures close magnetic circuit reducing leakage flux and solves the problem of insulating the high voltage winding from the core. The purpose of laminating the core is to reduce the eddy current loss and use of silicon steel reduces the hysteresis loss. The windings placed on the core are of concentric nature and are termed as concentric or cylindrical windings. The thickness of laminations is usually 0.14, 0.28, 0.3 or 0.35 mm. The laminations have a thin coating of insulating enamel liquid. Normally transformer windings have circular coils. To have more core area, stepped core is used instead of square or rectangular core. A 2 step core is termed as cruciform core. When transformers work on load, they become heated up due to copper and iron losses. Therefore proper cooling system is designed to keep the temperature at permissible level.

2.5 Shell Type Construction

The construction of shell type transformer is shown in Fig.2.3. The central limb is called the core upon which both the windings are placed. The outer two limbs form the yoke provide low reluctance flux path.

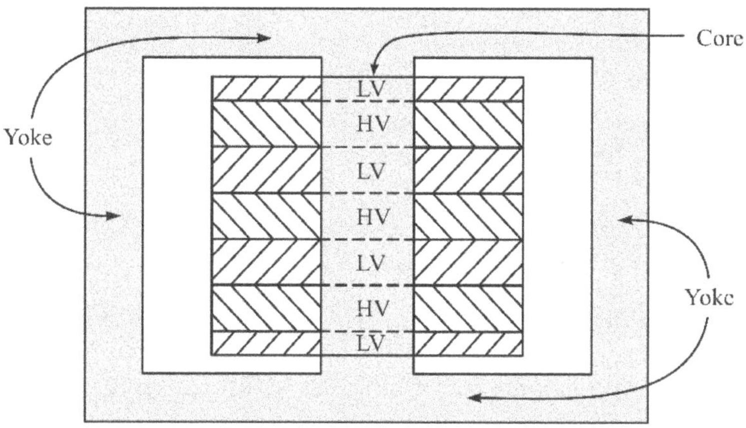

Figure 2.3 Single-phase Shell Type Transformer

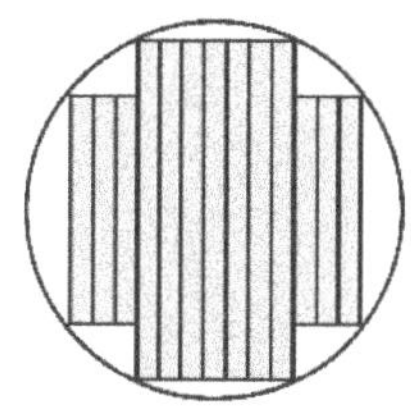

Figure 2.4 Section of Laminated Core (cruciform section)

The sectional area of the yoke is half of the sectional area of the central limb upon which both the windings are placed. In sandwich winding arrangements both LV & HV windings are divided into equal number of sections in the form of discs. One section of the LV winding is further divided into two equal sections. One of the sections is placed at the bottom of central core and then alternately HV and LV winding sections are placed. At the top of the central limbs the other half part of LV winding section is placed as shown in the Fig.2.3. This construction reduces leakage flux and provides mechanical strength to the windings. Sandwich windings are also termed as disc windings. Laminated core section is shown in Fig.2.4.

2.6 Types of Transformer Stampings

The common types of stampings (laminations) used in construction of transformers are in shape of L, U & I, E & I are shown in Fig.2.5.

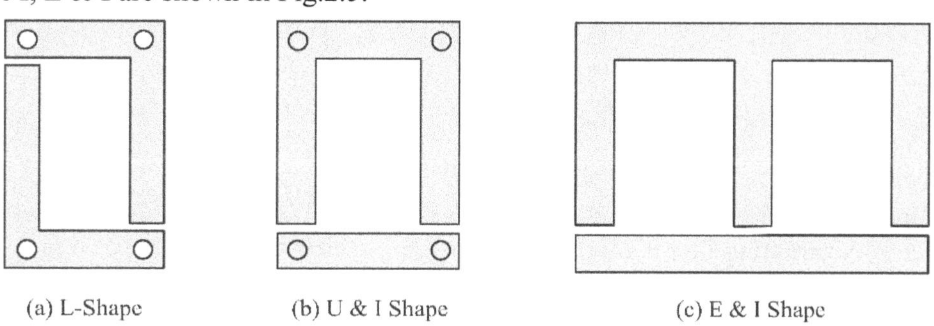

Figure 2.5 Transformer Stampings

The joints are butted together to make the laminated core structure firm.

2.7 EMF Equation

Consider a single-phase transformer having N_1 and N_2 turns of primary and secondary windings. Let there be a sinusoidally varying flux in the core and its value at any instant be

$$\phi(t) = \phi_m \sin \omega t \qquad .....(2.1)$$

The instantaneous value of induced emf in the primary winding would be

$$e_1(t) = -N_1 \frac{d\phi}{dt}(t) \qquad .....(2.2)$$

$$= -N_1 \frac{d}{dt}[\phi_m \sin \omega t]$$

$$= -N_1 \phi_m \omega \cos \omega t$$

$$= N_1 \phi_m \omega \sin\left(\omega t - \frac{\pi}{2}\right) \qquad .....(2.3)$$

Comparison of equations 2.1 and 2.3 shows that induced emf is also sinusoidal in nature and lags the flux by 90°.

In equation 2.3 the peak value of emf in primary winding is

$$E_{1(peak)} = N_1 \phi_m \omega \qquad .....(2.4)$$

$$= N_1 \phi_m 2\pi f \qquad .....(2.5)$$

where $\omega = 2\pi f$

The RMS value of induced emf in primary

$$E_1 = \frac{E_1(peak)}{\sqrt{2}} = \frac{N_1.\phi_m.2\pi f}{\sqrt{2}}$$

$$= 4.44 f \phi_m N_1 \qquad .....(2.6)$$

By similar procedure, we can get the RMS value of induced emf in the secondary winding as

$$E_2 = 4.44 f \phi_m N_2 \qquad .....(2.7)$$

Dividing equation 2.7 by equation 2.6, we get

$$\frac{E_2}{E_1} = \frac{4.44 f\phi_m N_2}{4.44 f\phi_m N_1} = \frac{N_2}{N_1} = k$$

where k is termed as the ratio of voltage transformation.

An important point to be noted is that when an ac supply is given to the primary winding of a transformer, an alternating flux is produced in the core, which lags the supply voltage by 90° as shown in Fig.2.6. Now the flux produced in the core induces an emf in the primary winding which lags the flux by 90°. Hence induced emf of the primary winding opposes the supply voltage. This is in confirmity with the Lenz's Law.

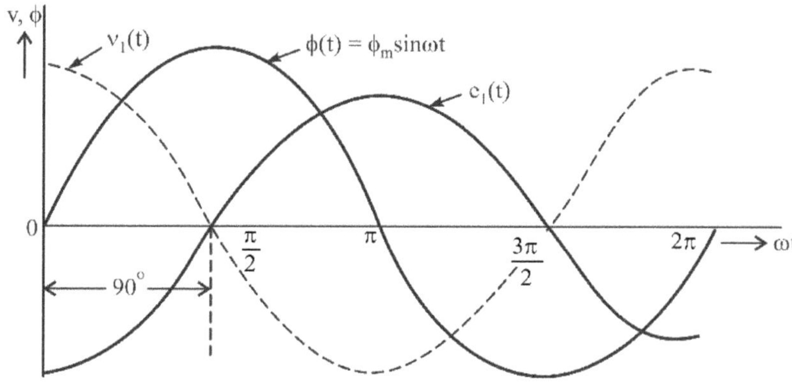

Figure 2.6 Diagram representing that induced emf wave $e_1(t)$ lags the flux wave by 90° and opposes $v_1(t)$

Example 2.1: A single-phase 6600/220 V, 50 Hz core type transformer has an effective core section of 0.05 m^2 and maximum flux density is 1.2 T. Determine the number of turns on HV and LV sides.

Solution:

EMF equation is

$$E_1 = 4.44\, f\, N_1\, \phi_m \text{ (V)} \qquad \text{.....(1)}$$

where $E_1 = 6600$ volt

$f = 50$ Hz

ϕ_m = core section × Max^m flux density

$= 0.05 \times 1.2 = 0.06$ Wb

From equation (1),

$$N_1 = \frac{E_1}{4.44 f \phi_m}$$

$$= \frac{6600}{4.44 \times 50 \times 0.06} = 496 \text{ turns}$$

The relationship between N_1 and N_2 is

$$\frac{E_1}{E_2} = \frac{N_1}{N_2} \qquad \text{..... (2)}$$

$$\therefore \qquad N_2 = \frac{E_2}{E_1}.N_1$$

$$= \frac{220}{6600} \times 496 = 16.53$$

Ans: Say 17 turns

Example 2.2: A 10 kVA, single-phase transformer has 200 turns in one winding and 100 turns in the other winding. The frequency of the supply is 50 Hz. If the induced emf in the primary side (when the transformer is used as a step-down transformer) is 250 V, find the value of maximum flux density if the core area is 64×10^{-4} m^2.

Solution: The induced emf E = 4.44 ϕ_mfN, where ϕ_m = maximum value of the flux in the core, f = input supply frequency and N = number of turns.

Here E = 250 V, ϕ_m = ?, f = 50 Hz and N = 200 turns. Hence

$$\phi_m = \frac{250}{4.44 \times 50 \times 200} = 5.6\ 10^{-3}\ \text{Wb}$$

Then the maximum flux density is

$$B_{max} = \frac{5.6 \times 10^{-3}}{64 \times 10^{-4}} = 0.875\ \text{Wb/m}^2$$

2.8 Equivalent Circuit and Phasor Diagram on No Load

A transformer has two windings named as primary winding and secondary winding. Each of the winding possesses some definite value of resistance say R_1 and R_2. With the transformer core, three types of fluxes are linked. In Fig.2.7 transformer mutual and leakages fluxes are shown. The flux which links with both the windings is called mutual or useful flux. The flux that links with primary winding is termed as primary winding leakage flux and the flux that links with the secondary winding only is called secondary winding leakage flux. The leakage fluxes are proportional to primary and secondary currents. On no load there would be no leakage flux on secondary side.

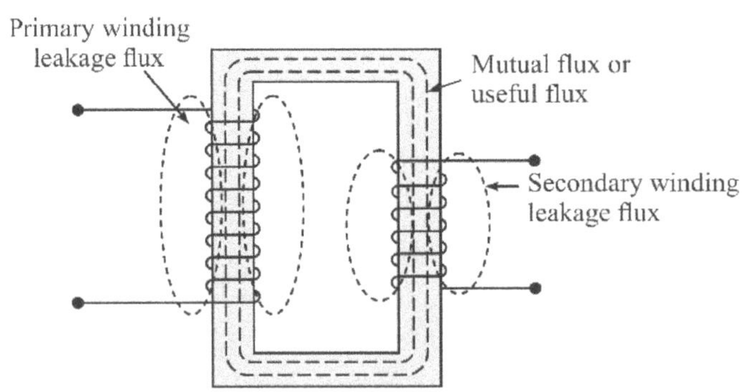

Figure 2.7 Transformer with mutual and Leakage Fluxes

The representation of a transformer by an equivalent circuit helps in determining the performance of the transformer. Let x_1 and x_2 be the leakage reactance's of primary and secondary windings. In Fig.2.8 a transformer circuit showing primary and secondary resistances and leakage reactance's are shown externally.

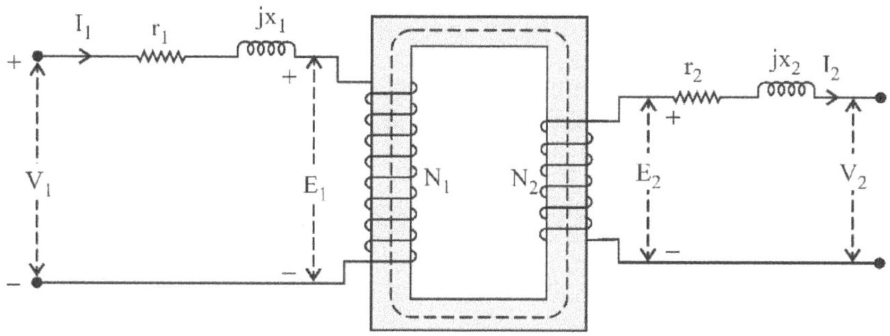

Figure 2.8 Transformer circuit showing windings (arrows mark as instantaneous directions) resistances and leakage reactance's external to the windings

Under no load condition the secondary side is open circuited and no current flows in the secondary winding. Only a very small current flows in the primary winding called as no load current (I_o). This current consists of following two components:

(a) I_μ = Magnetizing component of the no load current which lags behind the applied voltage V_1 by 90° and its function is to produce an alternating magnetic flux in the core.

(b) I_c = Core loss component of the no load current which supplies iron (Hysteresis and Eddy current) losses in the transformer core. This component is in phase with V_1 and is also known as active or working component.

Further the transformer winding resistances and leakage reactance's have negligible role under open circuit condition. Hence transformer circuit of Fig.2.8 reduces to a very simple equivalent circuit as shown in Fig.2.9. The magnetizing impedance $Z_o = \frac{V_1}{I_o}$ may be represented by a core loss resistance R_o in parallel with a magnetizing reactance X_o.

Based on the above analysis, the phasor diagram of the transformer under no load condition shall be as shown in Fig.2.10. The core loss component of current I_c is plotted along V_1, as it is in phase with it. The magnetizing component I_μ is plotted 90° lagging to V_1. The vector sum of I_c and I_μ gives I_o, the no load current. Angle ϕ_o between I_c and I_o is power factor angle on no load. Its value is normally 82° to 86°. The angles α between I_o and I_μ is called angle of hysteresis advance.

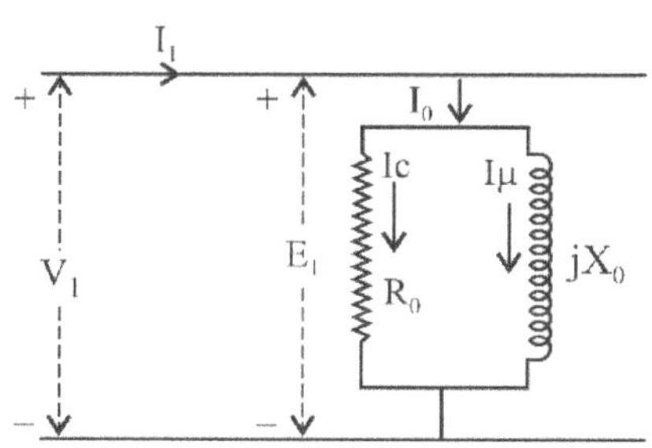

Figure 2.9 Equivalent circuit under no load condition

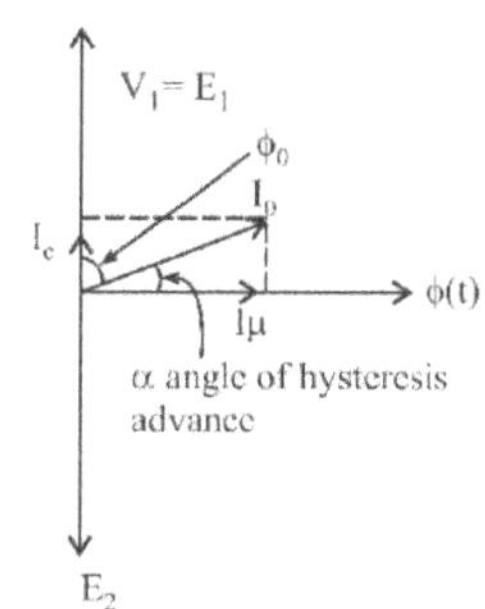

Figure 2.10 Phasor diagram of transformer under no load condition

2.9 Magnetizing Current v/s Magnetizing Inrush Current

Magnetizing current is that component of the no load current which establishes the sinusoidal core flux in the transformer. It is RMS value and is present in steady state operation.

The magnetizing inrush current depends upon the instant of supply cycle when the switch K is switched on in Fig.2.11.

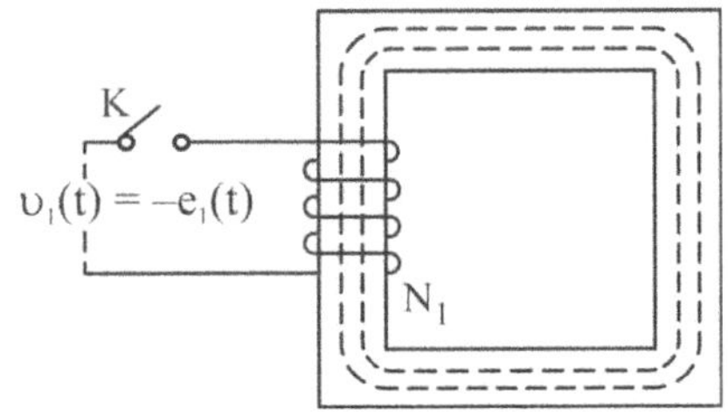

Figure 2.11 Inrush Current

The instantaneous value of applied voltage to the primary is:

$$V_1(t) = V_{1m} \cos(\omega t + \theta) \quad .....(1)$$

where V_{1m} = peak value of applied voltage

θ = Angular displacement

In magnitude

$$V_1(t) = e_1(t) = N_1 \frac{d\phi(t)}{dt} \quad .....(2)$$

From equations 1 and 2

$$N_1 \frac{d\phi(t)}{dt} = V_{1m} \cos(\omega t + \theta) \quad .....(3)$$

Integrating both sides

$$\phi(t) = \frac{V_{1m}}{N_1\omega} \sin(\omega t + \theta) + k \quad .....(4)$$

At t = 0

$$\phi(t) = \frac{V_{1m}}{N_1\omega} \sin\theta + k \quad .....(5)$$

Flux linkages at t = 0

$$\psi(t) = N_1\, \phi(t) = \frac{V_{1m}}{\omega} \sin\theta + k\, N_1 \quad .....(6)$$

Ignoring any residual flux, at t = 0, $\psi(t) = 0$, so $\frac{V_{1m}}{\omega} \sin\theta + kN_1 = 0$

$$\therefore \quad k = -\frac{V_{1m}}{N_1\omega} \sin\theta \quad .....(7)$$

Equation 4 gives

$$\phi(t) = \frac{V_{1m}}{N_1\omega} \sin(\omega t + \theta) - \frac{V_{1m}}{N_1\omega} \sin\theta$$

$$= \frac{V_{1m}}{N_1\omega} [\sin(\omega t + \theta) - \sin\theta] \quad ....(8)$$

Case 1: $\theta = 0$

(a) equation (1) & (8) give $v_1(t) = V_{1m}\cos\omega t$ and $\phi(t) = \frac{V_{1m}}{N_1\omega}\sin\omega t$ at t = 0, the voltage is passing through its maximum and at $t \geq 0$, the flux produced is sinusoidal without any transient.

Case 2: $\theta = \frac{\pi}{2}$

$v_1(t) = -V_{1m}\sin\omega t$, the supply voltage is passing through zero when the switch is closed. $\phi(t) = \frac{V_{1m}}{N_1\omega}[\cos\omega t - 1]$, showing a dc component of flux.

Case 3: $\omega t = \pi$

Equation 8 gives, $\phi(t) = \frac{V_{1m}}{N_1\omega}[-\sin\theta - \sin\theta] = -\frac{2V_{1m}}{N_1\omega}\sin\theta$

The flux is double the maximum value obtained in case 1 as shown in Fig. 2.12. Immediately it would cause a very high magnetizing inrush current which shall decay to zero due to winding resistance.

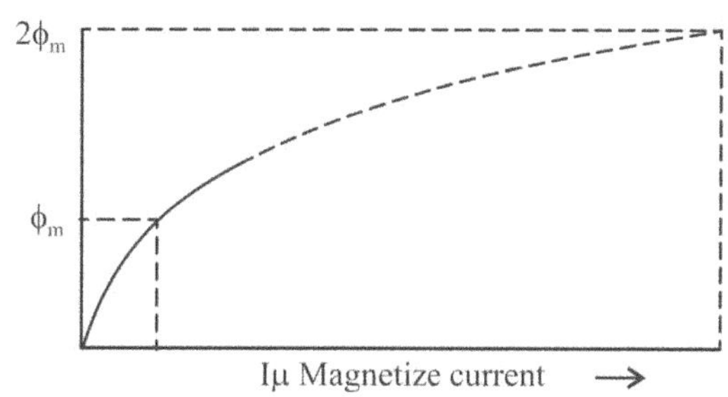

Figure 2.12 Magnetizing current

The following points are worth noting:

(i) For obtaining higher efficiency and good voltage regulation, I_μ and I_c both must be small.

(ii) I_μ is kept small by having a completely closed magnetic circuit of high permeability.

(iii) I_c is kept small by using low loss steel and working at suitable value of low flux density.

(iv) I_o is much smaller than the full load current (may be 2 - 6% of full load current).

(v) No load input power $W_o = V_1 I_o \cos\phi_o$, where $\cos\phi_o = \frac{I_c}{I_o}$ is power factor under no load condition.

(vi) Since I_o is very small, hence winding cu-loss on no load is negligibly small. Hence no load power input W_o represents iron losses of the transformer.

(vii) Since the no load component vector I_o is displaced due to the core loss, the angle α between I_μ and I_o is called the angle of hysteresis advance.

(viii) Though by using low flux density, there is a reduction in iron loss, but the weight of iron in transformer increases, making it bulky and costly. Hence a compromise is made between the two.

(ix) The values of R_o and X_o are given in Fig.2.9.

$$R_o = \frac{V_1}{I_c}$$

$$X_o = \frac{V}{I_\mu}$$

2.10 Equivalent Circuit and Phasor Diagram on Load

A simple circuit diagram of a transformer showing various resistances and leakage reactances externally, along with magnetic coupling as shaded part in Fig.2.13. The magnetizing impedance is represented by the parallel branch having core loss resistance R_o and magnetizing reactance X_o.

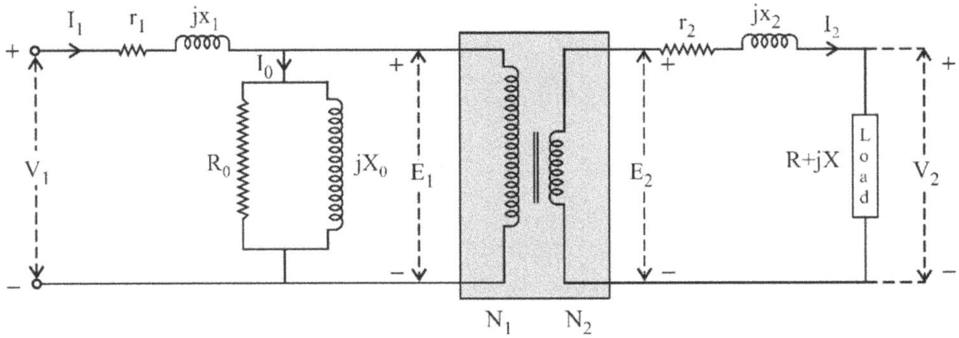

Figure 2.13 Simple Circuit Diagram of a Transformer

For analyzing the performance of the transformer, it is much more convenient to transfer all the quantities to either primary or secondary side. Such a circuit is termed as an "Equivalent Circuit" of the transformer. It is possible when secondary quantities are transferred to primary side or vice versa.

Let r_2' be the equivalent resistance of secondary when referred to primary side and I_2' be the equivalent current of I_2 when referred to primary side.

The copper loss in R_2 referred to primary side would be

$$I_2'^2 r_2' = I_2^2 r_2 \qquad \text{.....(1)}$$

$$\therefore \qquad r_2' = \left(\frac{I_2}{I_2'}\right)^2 r_2 \qquad \text{.....(2)}$$

The corresponding ampere turns referred to primary and secondary side would be

$$I_2' N_1 = I_2 N_2 \qquad \text{.....(3)}$$

$$\therefore \qquad \frac{I_2}{I_2'} = \frac{N_1}{N_2} \qquad \text{.....(4)}$$

Putting the value of $\frac{I_2}{I'_2}$ in equation 2 we get

$$r'_2 = \left(\frac{N_1}{N_2}\right)^2 r_2 \qquad .....(5)$$

Similarly, the equivalent secondary leakage reactance x_2 referred to primary side shall be

$$x'_2 = \left(\frac{N_1}{N_2}\right)^2 x_2 \qquad .....(6)$$

Load impedance referred to primary side

$$= \left(\frac{N_1}{N_2}\right)^2 (R + jx) \qquad .....(7)$$

Secondary induced emf E_2 referred to primary side is

$$E'_2 = \left(\frac{N_1}{N_2}\right) E_2 \qquad .....(8)$$

The terminal voltage v referred to primary side is

$$V'_2 = \left(\frac{N_1}{N_2}\right) V_2 \qquad .....(9)$$

Thus, the equivalent circuit of the transformer after referring all the quantities to the primary side is shown in Fig.2.14. This is known as "Exact Equivalent Circuit" of the transformer.

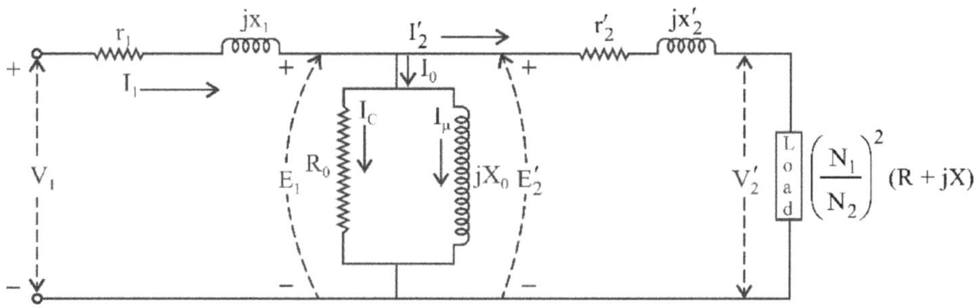

Figure 2.14 Exact Equivalent Circuit of the Transformer

This equivalent circuit can be further simplified by transferring the parallel branch near to the supply voltage as shown in Fig.2.15. This equivalent circuit is known as "Approximate Equivalent Circuit" and the analysis gets very much simplified.

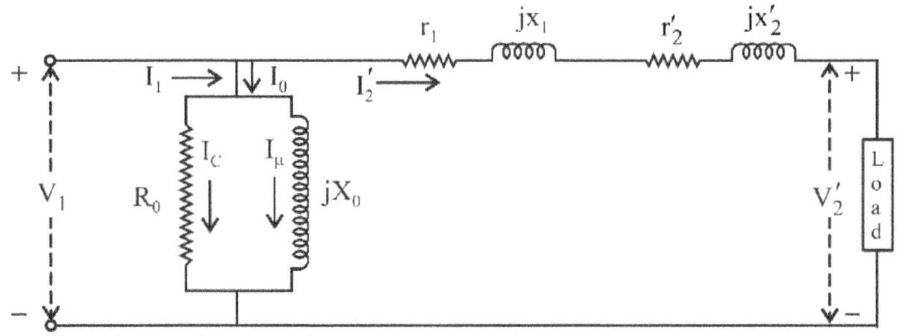

Figure 2.15 Approximate Equivalent Circuit of the Transformer

From Fig.2.12, the current I_2' is given by

$$I_2' = I_1 - I_o \qquad \text{.....(10)}$$

or

$$I_1 = I_2' + I_o \qquad \text{.....(11)}$$

It may be observed from the circuit of Fig.2.15 that the two resistances r_1 and r_2'; and two leakage reactance's x_1 and x_2' may be combined together to obtain a much-simplified approximate equivalent circuit as shown in Fig.2.16.

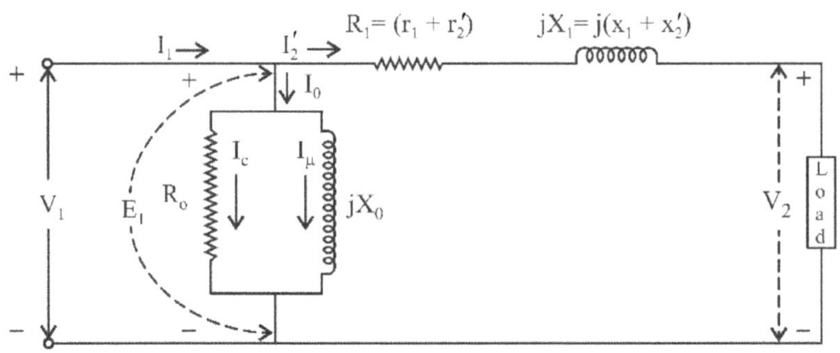

Figure 2.16 Simplified Approximate Equivalent Circuit of the Transformer

Total equivalent resistance referred to the primary side

$$R_1 = r_1 + r_2' \qquad \text{.....(12)}$$

Total equivalent leakage reactance referred to the primary side

$$X_1 = x_1 + x_2' \qquad \text{.....(13)}$$

When we change over from exact equivalent circuit to approximate equivalent circuit the following two errors are caused:

(a) The current through r_1 becomes I_2' in place of I_1. Current I_2' is somewhat less than I_1. This gives somewhat less copper loss in primary winding than the actual.

(b) The current I_o in circuit of Fig.2.16 is somewhat greater than that in circuit of Fig.2.14. Since I_o is due to V_1 in Fig.2.16 and due to E_1 in Fig.2.14. Here E_1 is less than V_1 due to voltage drop in $(r_1 + jx_1)$. Hence this gives somewhat greater loss in R_o than the actual.

Obviously the above two errors are of opposite nature and hence the results obtained by approximate equivalent circuit are fairly accurate.

To have a clear understanding of the phasor diagram of Fig.2.17 we take circuits of Fig.2.13 and 2.14 and begin from secondary side in the following steps.

- Draw V_2 and I_2 at lagging power factor angle ϕ_2.
- Draw $I_2 r_2$ drop parallel to I_2 vector and $j I_2 x_2$ drop 90° ahead of I_2 vector.
- The vector sum of V_2, $I_2 r_2$ and $j I_2 x_2$ drops give the secondary induced emf E_2.
- Draw vector E_1 opposite to OE_2 and its magnitude is given by the relation $E_1 = \frac{-N_1}{N_2} E_2$.
- Plot core loss component I_c along the vector $-E_1$ and I_μ perpendicular to it.
- The vector sum of I_c and I_μ gives the no load current I_o.

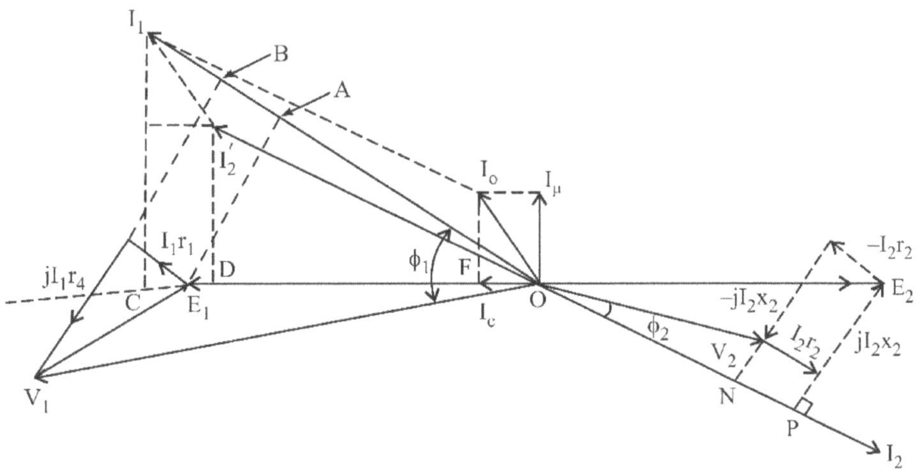

Figure 2.17 Phasor Diagram based on exact Equivalent Circuit of Fig.2.14

- Draw vector I_2' opposite to vector I_2 and its magnitude is given by the relation $I_2' = \frac{-N_2}{N_1} I_2$
- The vector sum of I_2' and I_o gives the primary input current I_1.
- Draw $I_1 r_1$ drop parallel to vector I_1 and $jI_1 x_1$ drop 90° ahead of I_1 vector.
- The vector sum of E_1, $I_1 r_1$ and $j I_1 x_1$ gives the primary supply voltage V_1.
- The phasor diagram of Fig.2.17 refers to the exact equivalent circuit of the transformer given in Fig.2.14.

2.11 Power Flow Analysis

From the phasor diagram of Fig.2.17, we can do the complete power flow analysis of the transformer.

Power input (P_i) to the transformer is:

$$P_i = V_1 I_1 \cos \phi_1 \quad .....(1)$$

where V_1 = Input voltage (V)

I_1 = Input current (A)

$\cos \phi_1$ = Power factor on the primary input side

Or $P_i = I_1 V_1 \cos \phi_1$

$$= I_1 OB = I_1(OA + AB) = I_1(OA + I_1 r_1)$$

$$= I_1 OA + I_1^2 r_1$$

$$= I_1 . OE_1 \cos \phi_{E_1}^{I_1} + I_1^2 r_1$$

Since OA is equal OE_1 cos of angle between E_1 and I_1. We may again express

$$P_i = E_1 . I_1 \cos \phi_{E_1}^{I_1} + I_1^2 r_1$$

$$= E_1 . OC + I_1^2 r_1$$

$$= E_1(OD + DC) + I_1^2 r_1$$

$$= E_1 . OD + E_1 . DC + I_1^2 r_1$$

$$= E_1 . I_2' \cos \phi_{E_1}^{I_2'} + E_1 . OF + I_1^2 r_1$$

$$= I_2' . E_1 \cos \phi_{E_1}^{I_2'} + E_1 . I_c + I_1^2 r_1$$

We have seen, $I_2' = \dfrac{-N_2}{N_1} I_2$, $E_1 = \dfrac{-N_1}{N_2} E_2$ and $\cos \phi_{E_2}^{I_2} = \cos \phi_{-E_1}^{-I_2'}$. The power input expression becomes,

$$P_i = \left(\frac{-N_2}{N_1} I_2\right)\left(\frac{-N_1}{N_2} E_2\right) \cos \phi_{E_2}^{I_2} + E_1 I_c + I_1^2 r_1$$

$$= E_2 I_2 \cos \phi_{E_2}^{I_2} + E_1 I_c + I_1^2 r_1$$

Power transferred to secondary side by electromagnetic induction plus transformer core loss plus primary winding copper loss.

Power transferred to secondary side

$$= E_2 I_2 \cos \phi_{E_2}^{I_2}$$

$$= I_2 . E_2 \cos \phi_2 = I_2 . OP$$

$= I_2 (ON + NP) = I_2.ON + I_2 NP$

$= I_2 . V_2 \cos \phi_2 + I_2^2 r_2$

$= V_2 I_2 \cos \phi_2 + I_2^2 r_2$

= Power output on secondary side + secondary winding copper loss

The power flow diagram of the transformer is given in Fig.2.18.

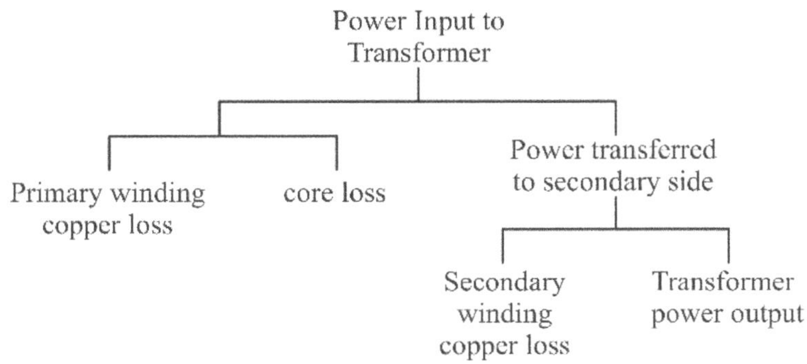

Figure 2.18 Power Flow Diagram

2.12 Per Unit (pu) Value

When quantities like resistance, reactance, copper loss, impedance, voltage, current etc. are expressed in terms of ohm, volt, ampere etc, they apply to a particular transformer. If these quantities be expressed in terms of per unit values, then they would be applicable for wide range of transformers and better comparison would be possible. Percentage resistance is defined as the resistance drop in volts at rated current as percentage of rated voltage

$$\% R = \frac{IR}{V} \times 100$$

$$\text{Per unit } R = \frac{IR}{V}$$

where R is ohmic value of resistance (Ω)

I is rated current (A)

V is rated voltage (V)

Similarly, reactance and impedance may be expressed as

$$\% X = \frac{IX}{V} \times 100$$

where X is ohmic value of reactance.

$$\% Z = \frac{IZ}{V} \times 100$$

when Z is ohmic value of impedance.

The short circuit current of a transformer is given by:

$$I_{sc} = \frac{100}{\%\ \text{Impedance}} \times \text{Full load current}$$

2.13 Voltage Regulation

The change in secondary terminal voltage from full load to no load at rated power factor, expressed as percentage of the no load terminal voltage, is called percentage voltage regulation under the condition when rated primary applied voltage is kept constant.

$$\%\ \text{Regulation} = \frac{E_2 - V_2}{E_2} \times 100$$

where

E_2 = Secondary terminal voltage on no load when rated voltage is applied on primary side

V_2 = Secondary voltage when rated load at rated pf is applied

Voltage regulation gives us an idea of the way in which the secondary terminal voltage varies with the load. It depends upon the load current, internal impedance of the transformer and the load power factor.

Voltage regulation may be estimated by taking approximate equivalent circuit referred to the secondary side as shown in Fig.2.19.

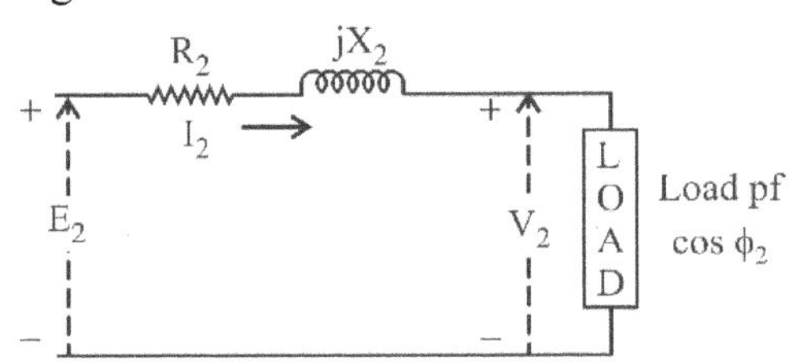

Figure 2.19 Equivalent Circuit referred to Secondary Side

R_2 = Total equivalent resistance referred to secondary side (Ω)

X_2 = Total equivalent leakage reactance referred to secondary side (Ω)

I_2 = Rated load current on secondary side (A)

$\cos \phi_2$ = Load power factor (say lagging)

The phasor diagram Fig.2.20 is drawn in the following way:

- Draw V_2 as reference vector
- Draw I_2 lagging V_2 by angle ϕ_2
- Draw $I_2 R_2$ drop parallel to I_2
- Draw jI_2X_2 drop 90° ahead of vector I_2

- The vector sum of V_2, $I_2 R_2$ and jI_2X_2 gives E_2.

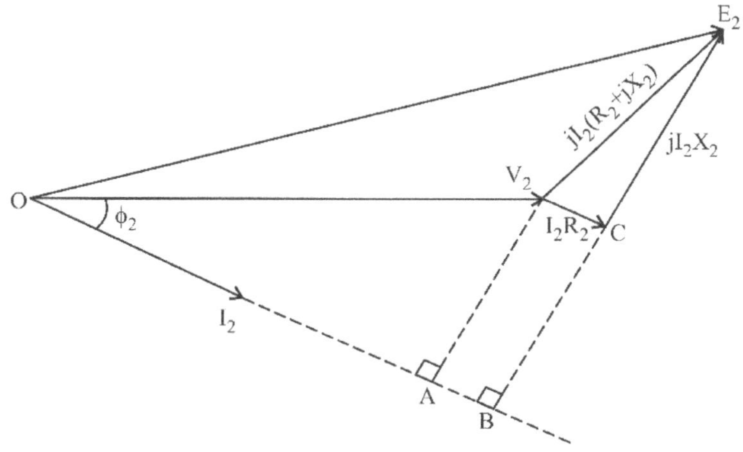

Figure 2.20 Phasor Diagram

Δ E_2OB is a right angle Δ, from this Δ we have

$$E_2 = \sqrt{OB^2 + E_2B^2}$$

$$= \sqrt{(OA + AB)^2 + (CB + E_2C)^2}$$

$$= \sqrt{(V_2 \cos\phi_2 + I_2R_2)^2 + (V_2 \sin\phi_2 + I_2X_2)^2}$$

Thus % Regulation $= \dfrac{E_2 - V_2}{E_2} \times 100$

The value of E_2 can also be obtained by drawing the phasor diagram to the scale.

Alternative Definition of Voltage Regulation

Sometimes in the definition of Regulation, the change in voltage is expressed with respect to V_2 instead of E_2. This is termed as up-voltage regulation.

Up-voltage Regulation $= \dfrac{E_2 - V_2}{V_2} \times 100\%$

The voltage regulation as already defined before is termed as Down-voltage regulation.

Down voltage regulation $= \dfrac{E_2 - V_2}{E_2} \times 100\%$

2.13.1 Voltage Regulation by Approximate Method

Consider the equivalent circuit and phasor diagram as given in Fig.2.19 and Fig.2.20. A small modification is made in this phasor diagram as explained and shown in Fig.2.21.

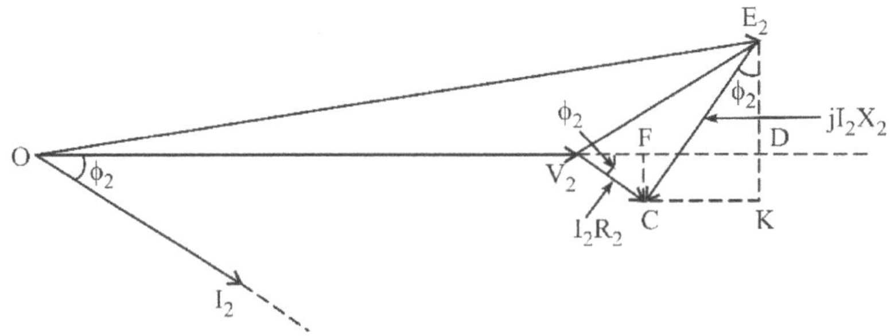

Figure 2.21 Phasor Diagram for determining regulation by Approx. method

- Extend line OV_2 to V_2D and draw the line CK parallel to this line.
- From point E_2 drop perpendicular on lines V_2D and CK meeting at D and K respectively.
- The no load secondary EMF E_2 may be expressed in j-notation as under:

$$E_2 = OD + j\,E_2D \quad .....(1)$$

$$= (OV_2 + V_2F + FD) + j(E_2K - DK)$$

$$= (OV_2 + V_2F + CK) + j(E_2K - FC)$$

$$= V_2 + I_2 R_2 \cos \phi_2 + I_2X_2 \sin \phi_2 + j\,(I_2X_2 \cos \phi_2 - I_2R_2 \sin \phi_2) \quad .....(2)$$

Since the leakage reactance of a transformer is quite small, the j component E_2D is much smaller than the horizontal component OD. So we may take in magnitude.

$$OE_2 \approx OD$$

$$\approx V_2 + I_2 R_2 \cos \phi_2 + I_2 X_2 \sin \phi_2 \quad .....(3)$$

The % regulation of the transformer is

$$\%\text{ Regulation} = \frac{E_2 - V_2}{E_2} \times 100$$

$$= \frac{I_2R_2}{E_2} \times 100 \cos \phi_2 + \frac{I_2X_2}{E_2} \times 100 \sin \phi_2$$

$$= \%\,R_2 \cos \phi_2 + \%\,X_2 \sin\phi_2 \quad .....(4)$$

where $\%\text{ resistance} = \frac{I_2R_2}{E_2} \times 100 = V_r$ say ....(5)

$\%\text{ leakage reactance} = \frac{I_2X_2}{E_2} \times 100 = V_x$ say .....(6)

$$\%\text{ Regulation} = V_r \cos \phi_2 + V_x \sin \phi_2 \quad .....(7)$$

where V_r and V_x are the total percentage equivalent resistance and leakage reactance of the transformer respectively referred to the secondary side.

At leading Power Factor

% Regulation = $V_r \cos\phi_2 - V_x \sin\phi_2$ .....(8)

Power Factor for Zero Regulation

From equation 4, the percentage regulation shall be zero,

when $\% R_2 \cos\phi_2 + \% X_2 \sin\phi_2 = 0$ .....(9)

or $\tan\phi_2 = -\dfrac{R_2}{X_2}$ .....(10)

The –ve sign indicates that the pf shall be leading for zero regulation.

Power Factor for Maximum Regulation and Corresponding value of Maximum Regulation

From equation 4 Percent Regulation is

$$\%\text{Reg} = V_r \cos\phi_2 + V_x \sin\phi_2 \quad .....(11)$$

For maximum value of regulation with reference to power factor angle ϕ_2,

$$\frac{d(\%\text{Reg})}{d\phi_2} = V_r \sin\phi_2 - V_x \cos\phi_2 = 0 \quad .....(12)$$

$$\therefore \tan\phi_2 = \frac{V_x}{V_r} = \frac{X_2}{R_2} \quad .....(13)$$

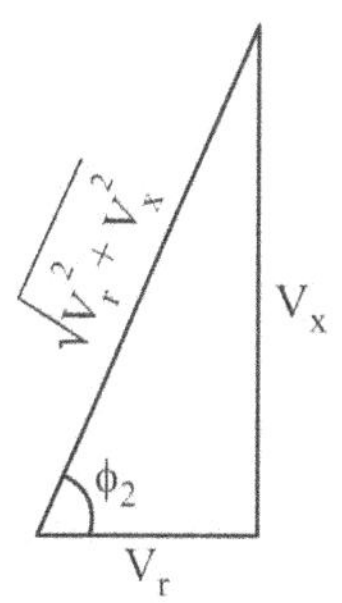

Figure.2.22

Draw the impedance triangle in terms of V_r and V_x as shown in Fig.2.22.

$$\cos\phi_2 = \frac{V_r}{\sqrt{V_r^2 + V_x^2}} \quad .....(14)$$

$$\sin\phi_2 = \frac{V_x}{\sqrt{V_r^2 + V_x^2}} \quad .....(15)$$

Substituting values of $\cos\phi_2$ and $\sin\phi_2$ in eq.(11) we get the maximum value of % regulation

$$\%\text{Reg} = V_r \frac{V_r}{\sqrt{V_r^2 + V_x^2}} + V_x \frac{V_x}{\sqrt{V_r^2 + V_x^2}}$$

$$= \sqrt{V_r^2 + V_x^2}$$

$$= \frac{I_2\sqrt{R_2^2 + X_2^2}}{E_2} \times 100 \quad ....(16)$$

2.14 Losses in a Transformer

There are two types of losses in a transformer. They are:

1. Iron or core loss
2. Copper loss

Since transformer is static machine, so there are no friction and windage losses.

Iron loss

Iron loss is produced in the transformer core by the alternating flux. It consists of hysteresis loss and eddy current loss. These losses are given by following expressions.

$$\text{Hysteresis loss} = K_h B_m^n f$$

$$\text{Eddy current loss} = K_e B_m^2 f^2 t^2$$

where K_h = A constant for hysteresis loss

K_e = A constant for eddy current loss

B_m = Peak value of flux density in the core (T)

f = frequency of supply (Hz)

t = Thickness of the lamination (m)

n = Steinmetz constant (1.5 – 2.0)

The value of B_m practically remains constant in a transformer from no load to full load. Frequency & thickness of laminations are also constant. Hence iron loss is considered as constant loss at all loads. The iron loss is determined by open circuit test on a transformer.

The iron loss is given by

$$W_i = \text{Hysteresis loss} + \text{Eddy current loss}$$

$$= K_h B_m^n f + K_e B_m^2 f^2 t^2$$

Copper Loss

It is due to the ohmic resistances of the transformer windings. Total copper loss

$$W_{cu} = \text{Loss in primary windings} + \text{Loss in secondary windings}$$

$$= I_1^2 r_1 + I_2^2 r_2$$

$$= \left(I_2 \frac{N_2}{N_1}\right)^2 r_1 + I_2^2 r_2$$

$$= I_2^2 \left[r_2 + \left(\frac{N_2}{N_1}\right)^2 r_1\right]$$

$$= I_2^2 . R_2$$

where R_2 is total equivalent resistance referred to secondary side.

In terms of transformer KVA rating, we may express the copper losses in the following way.

(KVA output) is proportional to I_2 and copper loss is proportional to I_2^2. Hence copper loss be taken as proportional to (KVA output)^2. At half load, the copper loss shall be $\left(\frac{1}{2}\right)^2$ of full load Copper loss, i.e., one fourth of the full load copper loss.

Thus at "x" fraction of full load, the copper loss would be.

$W_{cu}(x) = x^2$. Full load copper loss

So, at $x = 3/4$, i.e., three-fourth full load, the copper loss shall be $\left(\frac{3}{4}\right)^2$ of full load copper loss $= \frac{9}{16}$ Full load copper loss.

2.15 Efficiency

The efficiency is defined as the ratio of output power to input power.

$$\text{Efficiency } (\eta) = \frac{\text{Output power}}{\text{Input power}} \qquad \text{.....(1)}$$

where output power $= V_2 I_2 \cos\theta_2$ (W)

Input power = output power + losses

Losses = W_i (W) Iron loss

and $I_2^2 R_2$ (W) copper loss

$$\therefore \qquad \text{Efficiency} = \eta = \frac{V_2 I_2 \cos\theta_2}{V_2 I_2 \cos\phi_2 + W_i + I_2^2 R_2} \qquad \text{.....(2)}$$

To find the load current I_2 at which the η shall be maximum,

$$\frac{d\eta}{dI_2} = \frac{\left(V_2 I_2 \cos\phi_2 + W_i + I_2^2 R_2\right)\frac{d}{dI_2}\left(V_2 I_2 \cos\phi_2\right) - \left(V_2 I_2 \cos\phi_2\right)\frac{d}{dI_2}\left(V_2 I_2 \cos\phi_2 + W_i + I_2^2 R_2\right)}{\left(V_2 I_2 \cos\phi_2 + W_i + I_2^2 R_2\right)^2} = 0$$

or $\left(V_2 I_2 \cos\phi_2 + W_i + I_2^2 R_2\right) \times V_2 \cos\phi_2 - \left(V_2 I_2 \cos\phi_2\right)\left(V_2 \cos\phi_2 + 0 + 2 I_2 R_2\right) = 0$

or $V_2 I_2 \cos\phi_2 + W_i + I_2^2 R_2 - V_2 I_2 \cos\phi_2 - 2 I_2^2 R_2 = 0$

$$\text{or} \qquad I_2^2 R_2 = W_i \qquad \text{.....(3)}$$

or $$I_2 = \sqrt{\frac{W_i}{R_2}} \qquad .....(4)$$

where W_i = constant Iron loss

R_2 = Total equivalent resistance referred to secondary side.

In case of distribution transformers load current I_2 for maximum efficiency is kept between 65-75% considering that the transformer will work on this load most of the time.

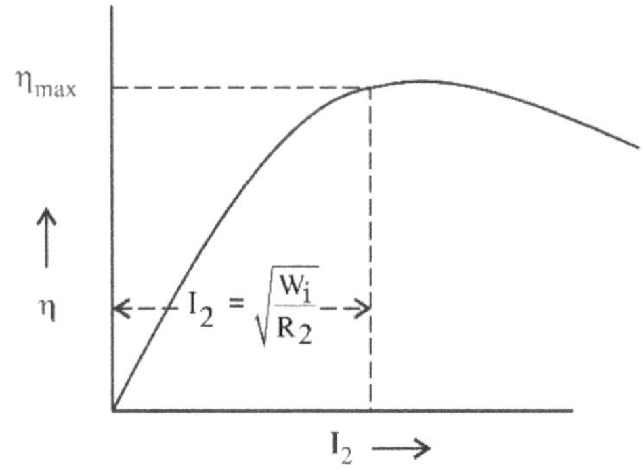

Figure 2.23 Curve between efficiency (η) and load current (I_2)

A curve is plotted between efficiency (η) and load current I_2. For value of load current $I_2 = \sqrt{\frac{W_i}{R_2}}$, the efficiency η_{max} shall be maximum as shown in Fig.2.23. If the transformer works at any other load, the efficiency shall be less than the maximum efficiency.

The equation (1) of η may alternatively be expressed as

$$\text{Efficiency } \eta = \frac{\text{Input} - \text{Losses}}{\text{Input}} \qquad .....(5)$$

$$= \left(1 - \frac{\text{Losses}}{\text{Input}}\right) \qquad .....(6)$$

Percentage efficiency would be

$$\%\eta = \left(1 - \frac{\text{Losses}}{\text{Input}}\right) \times 100 \qquad .....(7)$$

Example 2.3: A 5 kVA, single-phase 50 Hz, 1000/500 V transformer has the following equivalent circuit parameter values $r_1 = 0.1\ \Omega$, $x_1 = 0.3\,\Omega$; $r_2 = 0.004\,\Omega$, $x_2 = 0.012\ \Omega$. Find the equivalent resistance and reactance referred to the low voltage side. Also, find the full-load resistance drop in the low-voltage side (taken as secondary here).

Solution: We know that

$$R_2 = r_2 + r_1\left(\frac{N_2}{N_1}\right)^2$$

Here, $$\frac{N_1}{N_2} = \frac{1000}{500} = 2 \quad \text{or} \quad \frac{N_2}{N_1} = \frac{1}{2}$$

Therefore, $$R_2 = 0.004 + 0.1 \times \frac{1}{2^2} = 0.029\ \Omega$$

Similarly, $$X_2 = x_2 + x_1\left(\frac{N_2}{N_1}\right)^2 = 0.012 + 0.3 \times \frac{1}{2^2} = 0.087\ \Omega$$

Full load secondary current is

$$I_2 = \frac{5,000}{500} = 10\ \text{A}$$

Full-load resistance drop is

$$I_2R_2 = 10 \times 0.029 = 0.29\ \text{V}$$

Example 2.4: A 20 kVA, 2000 V/200 V, single-phase, 50 Hz transformer has a primary resistance of 1.5 Ω and reactance of 2 Ω. The secondary resistance and reactance are 0.015 Ω and 0.02 Ω respectively. The no-load current of transformer is 1 A at 0.2 lagging power factor. Determine.

(a) Equivalent resistance, reactance and impedance referred to primary

(b) Supply current

(c) Total copper loss

Solution: The turn ratio is given as

$$\frac{N_1}{N_2} = \frac{2000}{200} = 10$$

We also have: $r_1 = 1.5\ \Omega$, $x_1 = 2\,\Omega$. Also, $r_2 = 0.015\ \Omega$, $x_2 = 0.02\ \Omega$

(a) The equivalent resistance is

$$R_1 = r_1 + r_2\left(\frac{N_1}{N_2}\right)^2 = 1.5 + 0.015 \times \left(\frac{2000}{200}\right)^2 = 3\ \Omega$$

$$X_1 = x_1 + x_2\left(\frac{N_1}{N_2}\right)^2 = 2 + 0.02 \times \left(\frac{2000}{200}\right)^2 = 4\ \Omega$$

Impedance referred to the primary side will be

$$\text{Impedance} = \sqrt{R_1^2 + X_1^2} = \sqrt{3^2 + 4^2} = 5\ \Omega$$

Full load current at primary will be

$$I_1' = \frac{KVA}{V_1} = \frac{20 \times 10^3}{2000} = 10A$$

(b) Taking the load power factor to be unity, we get the supply current as

$$I_1 = I_1' + I_0 = 10\ \angle 0^\circ + 1\ \angle 78.46^\circ = 10.246\ \angle 5.48^\circ$$

(c) Total copper loss $\quad P_{cu} = 10^2 \times 3 = 300$ W

Example 2.5: A 25 kVA, 2200/220 V, 50 Hz, single-phase transformer has the following parameters $r_1 = 1.2\ \Omega$, $x_1 = 6.0\ \Omega$, $R_o = 1000\ \Omega$, $X_o = 300\ \Omega$; $r_2 = 0.03\ \Omega$, $x_2 = 0.06\ \Omega$ (low-voltage side). Draw the equivalent circuit referred to low-voltage side for a 0.8 pf lagging load supplied at a 220 V terminal voltage.

Solution: We have the transformer equivalent circuit as shown in Fig.2.24. Here from the equivalent circuit, we see that the load voltage, $V_2 = 220$ V. Assuming its angle to be zero, the angle of the load current (i.e I_2) will be $\cos^{-1} 0.8 = 36.87$ lagging. Now with

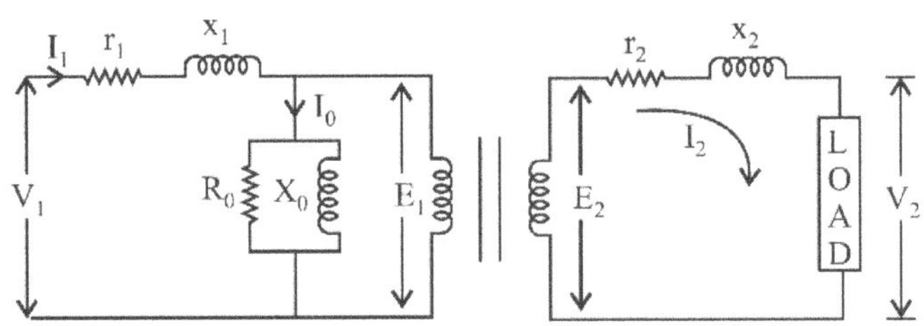

Figure 2.24 Equivalent Circuit

$$I_2 = \frac{25000}{220} = 113.64\ A$$

We have complete phasor expression of I_2, as

$$I_2 = 113.64\ \angle -36.87^\circ, \qquad V_2 = 220\ \angle 0^\circ$$

Now, in the low voltage (i.e., secondary side), we can get the induced emf, E_2, by adding the voltage drops I_2r_2 and I_2x_2. Thus, with $r_2 = 0.03\ \Omega$ and $x_2 = j0.06$, we have

$$E_2 = 220\ \angle 0^\circ + 113.64\ \angle -36.87^\circ \times 0.03 + 113.64\ \angle -36.87^\circ \times 0.06\ \angle 90^\circ$$

$$= 226.83 + j3.41$$

$$= 226.85\ \angle 0.86^\circ$$

Now we refer r_1, R_o and X_o towards the secondary (i.e., low-voltage) side to get their values as shown in Fig.2.25.

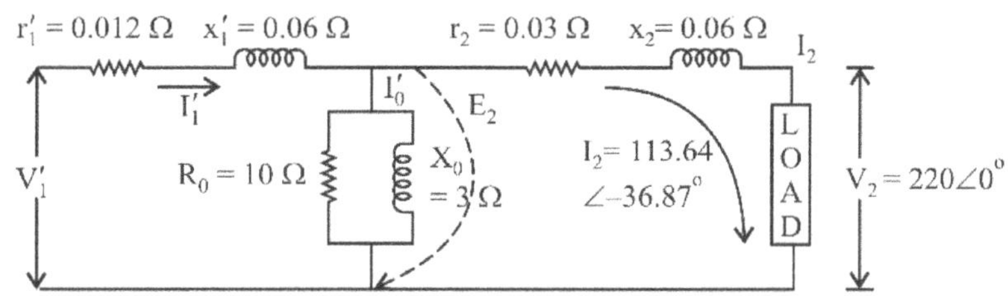

Figure 2.25 Values of variables in Fig.2.24

We need to find V_1'. For that we need to calculate, $I_1' = I_2 + I_0'$. Now I_0' (no-load current referred to the secondary) is

$$I_0' = \frac{E_2}{R_o'} + \frac{E_2}{jX_o'}$$

$$= \frac{22685\angle 0.86^\circ}{10} + \frac{226.85\angle 0.86^\circ}{3 < 90^\circ}$$

$$= 23.814 - j75.26$$

$$= 78.94 \angle -74.44$$

I_2 is already known as

$$I_2 = 113.64\ (0.8 - j0.6) = 90.912 - j68.18$$

and

$$I_1' = I_o' + I_2 = 114.73 - j143.45 = 183.68\angle -51.35$$

And finally

$$V_1 = E_2 + I_1'(0.012 + j0.06) = 226.85\angle 0.86^\circ + 183.68\angle -51.35^\circ \times 0.061\angle 78.69^\circ$$

Thus $V_1' = 237\text{V}$

2.16 All Day Efficiency

The distribution transformers remain connected to the supply all the 24 hours of the day irrespective of their load cycle. Core losses take place all the time continuously. The copper losses vary proportional to the square of load current. All day efficiency $\eta_{\text{all day}}$ is energy output in 24 hours to energy input in 24 hours.

$$\eta_{\text{all day}} = \frac{\text{Energy output in 24 hours}}{\text{Energy input in 24 hours}}$$

$$= \frac{\text{Energy output in 24 hours}}{\text{Energy output in 24 hours} + \text{copper loss in 24 hours} + \text{Iron loss in 24 hours}}$$

On general basis the distribution transformers are loaded 65-75% of their rating in 24 hours. Hence, they are so designed that their maximum efficiency be at 65-75% of the full load. It means Iron loss will be equal to copper loss at 65-75% of full load.

The Iron loss W_i (kW) remains constant throughout the day. If the copper loss be W_{cu} (kW) then copper loss at x fraction of full load will be $= x^2 . W_{cu}$ (kW)

Say, a distribution transformer works at full load for 10 hours, at half load for 8 hours and at no load for 6 hours in a day. On full load copper loss is W_{cu} (kW) and constant Iron loss is W_i (kW). Determine the all-day efficiency if the rating is P kVA at unity power factor.

$$\text{Energy output in 24 hours} = P \times 10 + \frac{1}{2} P \times 8$$

$$= 14P \text{ (kWh)}$$

$$\text{Iron loss} = W_i \times 24 \text{ (kWh)}$$

Copper loss on full load for 10 hours

$$= W_{cu} \times 10 = 10W_{cu} \text{ (kWh)}$$

Copper loss on half full load for 8 hours

$$= x^2 \cdot W_{cu} \times 8$$

$$= \left(\frac{1}{2}\right)^2 W_{cu} \times 8$$

$$= 2\ W_{cu} \text{ (kWh)}$$

Copper loss on no load for 6 hours = 0

$\therefore$ total copper loss $= 10\ W_{cu} + 2\ W_{cu}$

$$= 12\ W_{cu}$$

$$\therefore \qquad \eta_{\text{all day}} = \frac{14P}{14P + 24W_i + 12W_{cu}} \text{ (pu)}$$

$$= \frac{14P}{14P + 24W_i + 12W_{cu}} \times 100 \text{ (\%)}$$

Example 2.6: A 2300/230 V, 500 kVA, 50 Hz distribution transformer has core loss of 3.0 kW at rated voltage and copper loss of 6 kW at full load. During the day it is loaded as under:

% load	0%	50%	80%	100%
Pf		0.8 lag	0.9 lag	1
Hours	4	6	5	9

Determine all day efficiency of the transformer.

Solution:

kW output = kVA × Pf

Energy output = kW × hours

Total energy output may be calculated as under:

% rated load	PF	kVA cosϕ	kW	Hours	Output kWh
0		0	0	4	0
50	0.8	0.5 × 500 × 0.8	200	6	1200
80	0.9	0.8 × 500 × 0.9	360	5	1800
100	1.0	1.0 × 500 × 1	500	9	4500
			Total energy output =		7500

Full load copper loss = 6 kW

Copper loss on no load = 0 kWh

Copper loss at 50% load = x^2 $W_{cu(fl)}$ × Hours

$= 0.5^2 \times 6 \times 6$ = 9 kWh

Copper loss at 80% load = $0.8^2 \times 6 \times 5$ = 19.2 kWh

Copper loss at full load = 1 × 6 × 9 = 54 kWh

Total copper loss **= 82.2 kWh**

Total Iron loss in 24 hours = 3 × 24 **= 72 kWh**

Total energy input

= Total energy output + copper loss + Iron loss

= 7500 + 82.2 + 72

= 7654.2 kWh

$$\eta_{\text{all day}} = \frac{7500}{7654.2} = 0.98 \text{ pu}$$

or **98%**

2.17 Tests on Transformers

The following important tests are performed on transformers:

1. Open circuit test
2. Short circuit test
3. Polarity test
4. Sumpner's back to back test
5. Test for separation of Hysteresis and eddy current losses

2.17.1 Open Circuit Test

The open circuit test is normally performed on low voltage (LV) side of the transformer. The circuit diagrams for performing the open circuit test are given in Fig.2.26. In this test we need a voltmeter, ammeter and a wattmeter. The other winding of the transformer is kept open circuited.

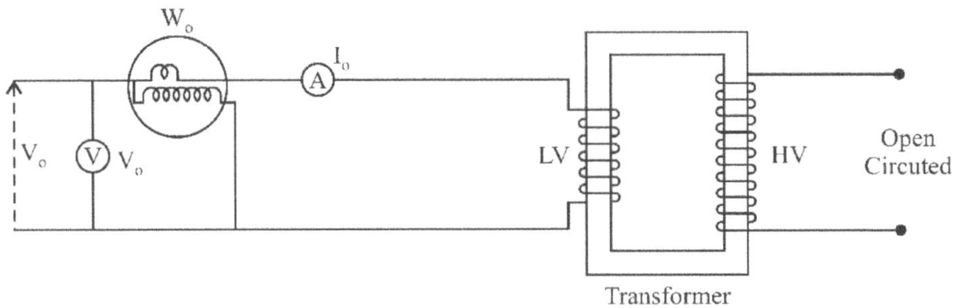

Figure 2.26 Circuit Diagram for open Circuit Test on Transformer

Rated voltage is applied on low voltage winding of the transformer. The readings of voltmeter, ammeter and wattmeter are

Let V_o = Rated applied voltage (V)

I_o = Ammeter reading on no load (A)

$W_o = W_i$ = Wattmeter reading on no load (W)

If $\cos\phi_0$ is pf under no load condition, then

$$V_o I_o \cos\phi_0 = W_o \quad .....(1)$$

or

$$\cos\phi_0 = \frac{W_o}{V_o I_o} \quad .....(2)$$

The equivalent circuit of the transformer under open circuit condition and the corresponding phasor diagram are shown in Fig.2.27 and 2.28 respectively.

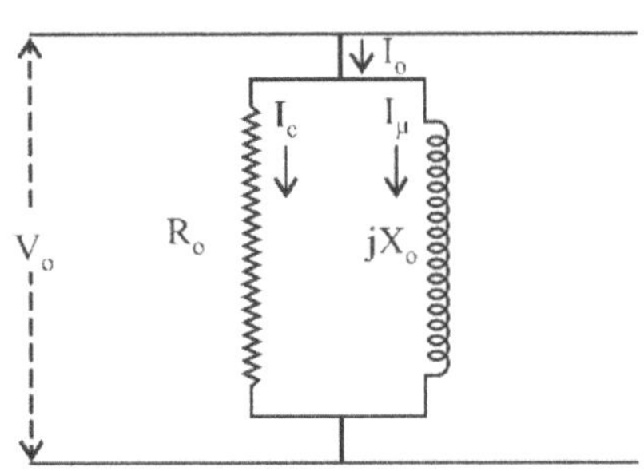

Figure 2.27 Equivalent Circuit

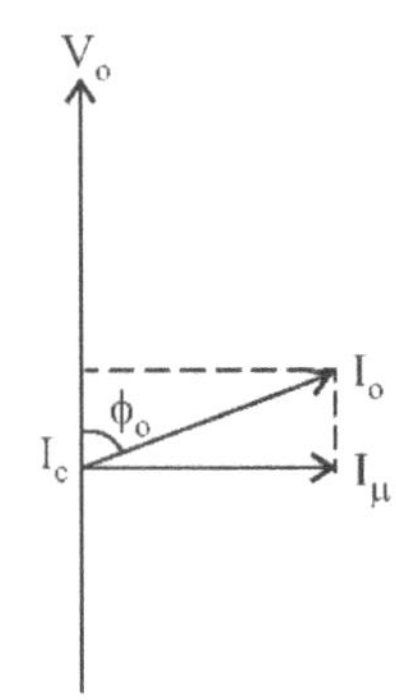

Figure 2.28 Phasor Diagram on Open Circuit

From phasor diagram of Fig.2.28

$$I_c = I_o \cos \phi_o \qquad \text{...... (3)}$$

$$I_\mu = I_o \sin \phi_o \qquad \text{....... (4)}$$

Hence,
$$R_o = \frac{V_o}{I_c} \qquad \text{...... (5)}$$

$$X_o = \frac{V_o}{I_\mu} \qquad \text{..... (6)}$$

At no load I_o is very small, so copper loss in transformer $I_0^2 R_1$ is also quite small. Hence entire input power on no load represents the iron loss of the transformer.

The reason of applying voltage on low voltage winding is due to practical consideration. Most of the distribution transformers are rated at 11/0.4 kV. Firstly 0.4 kV, 3 phase supply is commonly available in labs besides, being safer for performing this test. Further, no load current is very low on HV side. We work on LV side, so we get higher current values, hence more accurate metering on LV side.

2.17.2 Short Circuit Test

The short circuit test is normally performed on high voltage (HV) side of a transformer since we need only 2-6% rated voltage under short circuit condition to circulate rated current. This brings accuracy in measurement. The circuit diagram for performing the short circuit test is given in Fig.2.29. The LV winding of the transformer is short circuited.

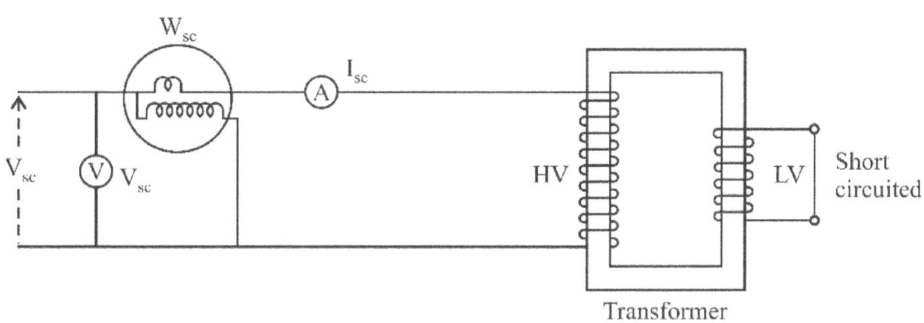

Figure 2.29 Circuit diagram for short circuit test on transformer

Now a variable low voltage is applied on the primary side through an auto transformer. Increase the supply voltage to V_{sc} to circulated $I_{sc} = I_{rated}$ on primary side. At this time take the wattmeter reading W_{sc}.

Let V_{sc} = Reduced applied voltage on HV side to circulate short circuit current I_{sc} equal to rated primary current I_1.

$I_{sc} = I_1$ = Rated primary current

W_{sc} = Wattmeter reading

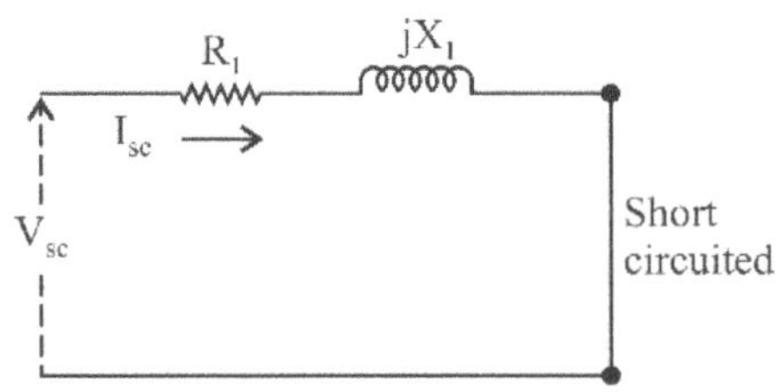

Figure 2.30 Equivalent Circuit

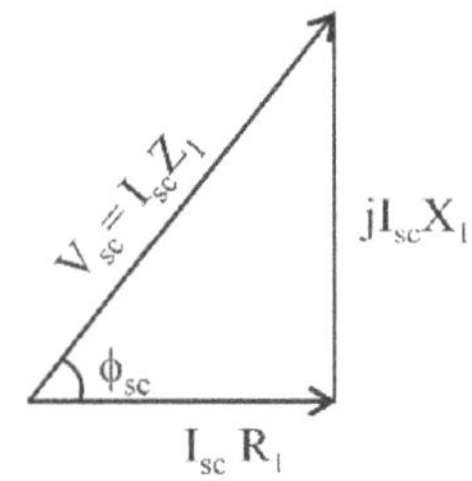

Figure 2.31 Phasor diagram under Short Circuit Condition

The equivalent circuit of the transformer under short circuit condition and corresponding phasor diagram are shown in Fig.2.30 and 2.31 respectively.

If $\cos \theta_{sc}$ is pf under short circuit condition, the

$$W_{sc} = V_{sc} I_{sc} \cos \phi_{sc} \quad .....(1)$$

$$\cos \phi_{sc} = \frac{W_{sc}}{V_{sc} I_{sc}} \quad .....(2)$$

Total impedance of the transformer

$$Z_1 = \frac{V_{sc}}{I_{sc}} \quad .....(3)$$

$$R_1 = Z_1 \cos \phi_{sc} \quad .....(4)$$

$$X_1 = Z_1 \sin \phi_{sc} \quad .....(5)$$

Since the applied voltage is very small under short circuit test, the flux setup in the core is also very small resulting in negligible iron loss. Hence the wattmeter reading W_{sc} gives total copper loss of the transformers.

2.17.3 Polarity Test

For proper connections of the transformer windings, it is essential to know correctly the instantaneous polarities of the secondary terminals with respect to the primary. Consider a two winding transformer as shown in Fig.2.32.

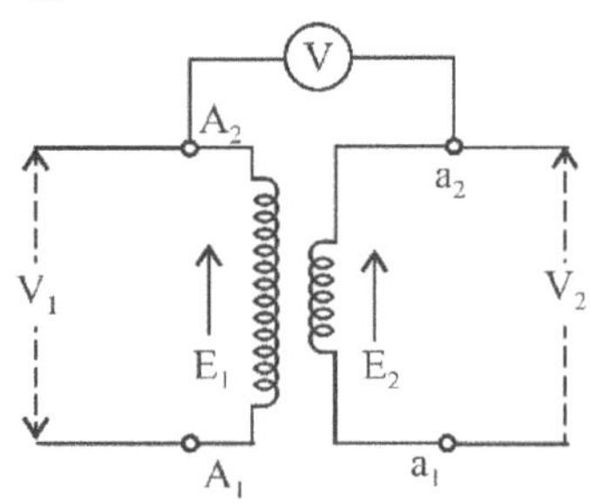

Figure 2.32 Polarity Test

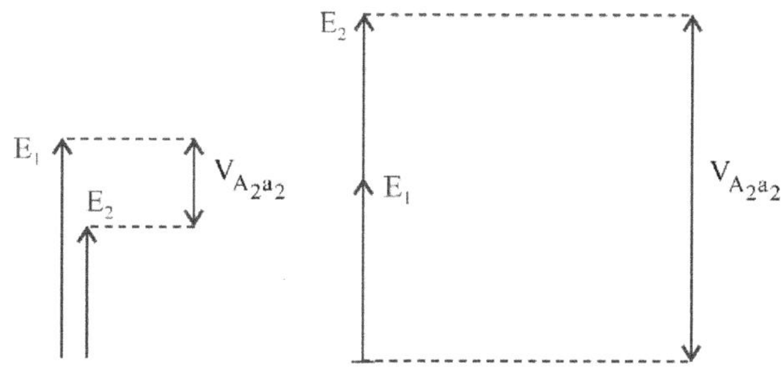

(a) correct polarity (b) Incorrect polarity

Figure 2.33 Phasors for Correct and Incorrect Polarity

On the primary side of the two-winding transformer, one terminal is positive with respect to the other one at any instant. At the same instant, one terminal of the secondary winding is also positive with respect to the other one. Polarity test is performed to determine the terminals having same instantaneous polarity.

In Fig.2.32 the terminals A_1 and a_1 are joined together and a voltmeter is placed across terminals A_2 and a_2. Apply a moderate voltage V_1 on primary side. A voltage V_2 will appear across terminals a_2 - a_1. For correct polarity the voltmeter V across $A_2 - a_2$ will show the difference E_1 and E_2 as shown in Fig.2.33(a). If the voltmeter V shows $E_1 + E_2$ voltage shown in Fig.2.33(b) the polarity is incorrect and the polarity marking must be interchanged.

Hence for correct polarity $V_{A_2a_2} < V_1$

Polarity test is essential for parallel operation of transformers, Sumpner's back to back test on transformers etc.

2.17.4 Sumpner's Back to Back Test

This test is performed on two identical transformers. In this test the great advantage is that both the transformers are fully loaded and can be made to run for hours consuming the energy to meet the losses only. This helps to determine the temperature rise of the transformer, which is of great practical importance. Besides the efficiency and regulation can also be determined. This test is performed in two parts.

In first part of the test the polarity of the transformer is tested as per circuit diagram of Fig.2.34.

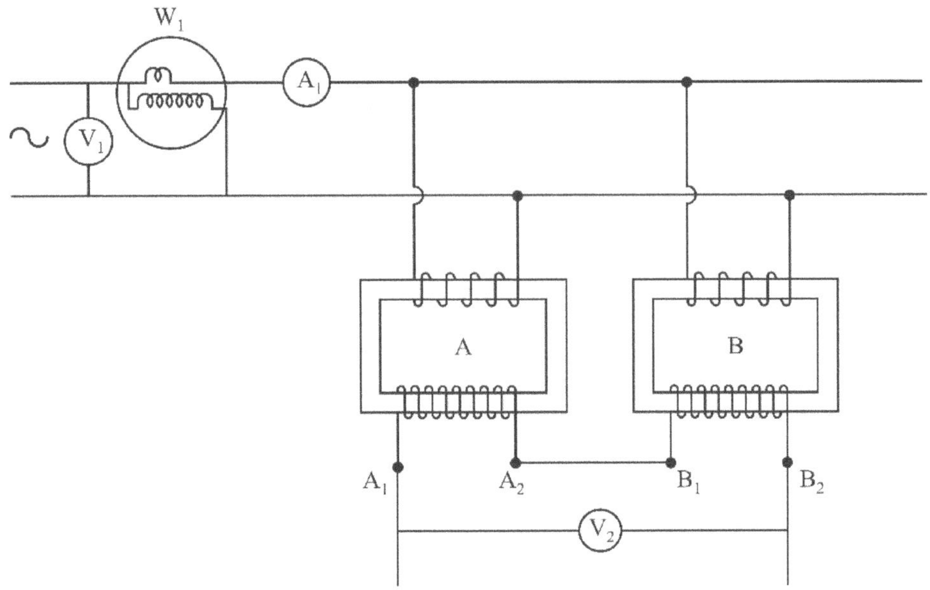

Figure 2.34 Part one of Sumpner's test to determine the correct polarity for back to back connection

Part One of the Test

- Connect a voltmeter V_1, ammeter A_1 and wattmeter W_1 in the supply line
- Connect two identical transformers A and B in parallel.
- On the secondary sides, the terminals are A_1 and A_2 for Transformer A and B_1 and B_2 for Transformer B.
- Connect terminals A_2-B_1 and place a voltmeter across terminals A_1–B_2
- Apply rated voltage V_1 on primary side and measure voltage V_2 across terminals A_1–B_2.
- The voltage between terminals A_1-B_2 shall be either zero or double of the secondary voltage.
- If the voltage is double, the connection is not back to back. For correct back to back connection, the voltage between terminals A_1–B_2 must be zero.
- Hence if the voltage between terminals A_1–B_2 is zero, the connection is correct and we proceed for part second of the test. If not correct, inter change any two terminals. Say, now the Fig.2.34 has correct polarity for back to back connection.

Part Second of the Test

The connections are made as shown in Fig.2.35. Connections of part one has to remain unchanged.

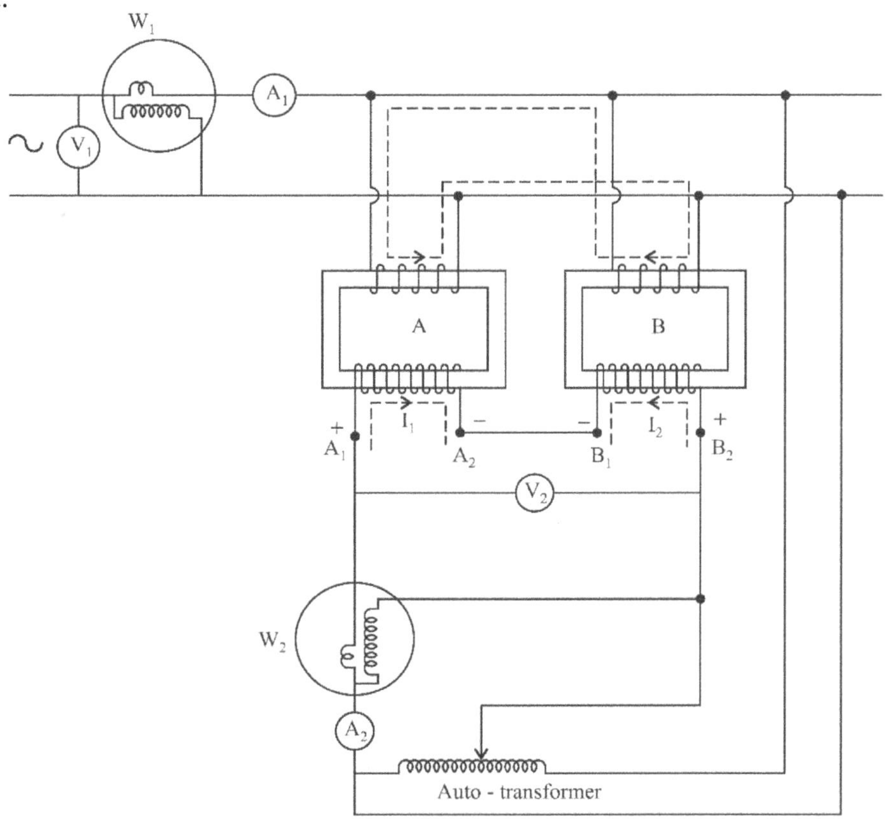

Figure 2.35 Complete circuit diagram for Sumpner's back to back test

- As a result of performing part one of the test, the two transformers are connected back to back and voltmeter V_2 reads zero volts.
- In part second of the test a reduced voltage is obtained from the same supply through an auto transformer to feed the transformers.
- An ammeter A_2 and a wattmeter W_2 are connected in the circuit as shown in Fig.2.35.
- By applying the voltage through the Autotransformer, a current I_2 starts flowing through the secondaries of both the transformers. Make this current equal to the rated secondary current.
- Take readings of voltmeter V_1, ammeter A_1 and wattmeter W_1 on the primary side and also the readings of voltmeter V_2, ammeter A_2 and wattmeter W_2.
- Sumpner's test combines open circuit test on primary side and short circuit test on secondary side.
- The readings thus obtained have the following relationships pertaining to open circuit test and short circuit test respectively.

$V_1 = V_o$

$I_1 = 2\, I_o$

$W_1 = 2\, W_o$

$V_2 = 2\, V_{sc}$

$I_2 = I_{sc}$

$W_2 = 2\, W_{sc}$

- With this data the equivalent circuit, regulation and efficiency can be easily determined as explained earlier.

2.18 Merits of Transformer

Transformers are extensively used due to the following merits.

(i) They are economical to transmit the electrical energy at higher voltages. The line losses get largely reduced at high voltage.

(ii) For distribution purpose, higher voltages can be stepped down with great ease.

(iii) The efficiency of transformers is very high.

(iv) Since transformer is a static device, it is very rugged and there is no wear and tear.

(v) The transformer can be designed for very high voltage, since it is easier to insulate the static windings of the transformer.

(vi) Very little attention is needed for its operation and maintenance.

(vii) They find wide application in instrumentation in the form of Current Transformer (CT) and Potential Transformer (PT).

Example 2.7: In a Sumpner's test on two identical single phase transformers rated 500 kVA, 11/0.4 kV, 50 Hz, the wattmeter reading on HV side is 6.0 kW and on LV side is 15 kW. Determine efficiency of each transformer on half load and 0.8 power factor lagging. What will be its maximum efficiency and at what percentage of load.

Solution:

Rating 500 kVA

The reading of wattmeter on hv side given iron losses of both the transformers, so iron loss of one transformer is given by.

$$W_i = \frac{1}{2} \times 6.0 = 3 \text{ kW}$$

The reading of wattmeter on lv side gives copper loss on full load for both the transformers.

S_o copper loss of one transformer on fill load is given by

$$W_{cu} = \frac{1}{2} \times 15 = 7.5 \text{ kW}$$

At half load copper loss will be

$$W_{cu\,(half\,load)} = \left(\frac{1}{2}\right)^2 \times 7.5$$

$$= 1.875 \text{ kW}$$

Output at half load at 0.8 pf

$$= \left(\frac{1}{2}\right) \times 500 \times 0.8 = 200 \text{ kW}$$

The efficiency at half load and 0.8 pf lagging would be,

$$\eta_{half\,load} = \frac{\text{Output power at half load}}{\text{Output power at half load} + \text{Copper loss at half load} + \text{Iron loss}}$$

$$= \frac{200}{200 + 1.875 + 3} \times 100$$

$$= \frac{200}{204.875} \times 100 = \mathbf{97.6\%\ Ans}$$

Maximum Efficiency

The load at which maximum efficiency occurs is given by the following relation

$$x^2 W_{cu} = W_i$$

or $x^2 \times 7.5 = 3$

$\therefore$ $x = \sqrt{\dfrac{3}{7.5}} = 0.632$ i.e., **63.2% of full load**

On this load output power at unity pf will be,

$= x \times 500 \times 1 = 0.632 \times 500 \times 1 = 316$ kW

Copper loss on this load $= x^2 \times 7.5$

$= 0.4 \times 7.5 = 3.0$ kW

Iron loss is constant, $W_i = 3$ kW

Maximum efficiency is given by

$$\eta_{max} = \frac{\text{Output power}}{\text{Out power} + \text{losses}}$$

$$= \frac{316}{316 + 3 + 3} \times 100$$

$$= \frac{316}{322} \times 100 = \mathbf{98.1\%}$$

2.19 Test for Separation of Hysteresis and Eddy Current Losses

The circuit for separation of losses is the same as that for open circuit test of the transformer. The iron loss is represented as,

W_i = Hysteresis loss + Eddy current loss

$$= K_h f B_m^n + K_e f^2 B_m^2 t^2 \quad .....(1)$$

where k_h and k_e are constants and t are the thickness

Primary voltage is given by

$$V_1 = 4 \cdot 44 f \phi_m N_1 \quad .....(2)$$

$$= 4 \cdot 44 f A_i B_m N_1 \quad .....(3)$$

or $$\frac{V_1}{f} = 4.44 A_i B_m N_1 \quad .....(4)$$

If the ratio $\frac{V_1}{f}$ is kept constant, then B_m is also constant since A_i & N_1 are already constant, A_i being the cross-sectional area of the core.

For this constant B_m, the Equation (1) for iron loss be written as

$$W_i = K_h' f + K_e' f^2 \quad .....(5)$$

where $$K_h' = K_h B_m^n \quad .....(6)$$

$$K_e' = K_e B_m^2 t^2 \quad .....(7)$$

Equation 5 may be written as

$$\frac{W_i}{f} = K_h' + K_e' f \quad ....(9)$$

The voltage and frequency are varied simultaneously keeping V/f ratio constant. The reading of wattmeter $W_o = W_i$ on no load test, taking reading of wattmeter for different frequency make the table as under

Table 2.1 Observation Table

S.No.	Frequency f	Watt meter $W_o = W_i$	Ratio $\frac{W_i}{f}$

Plot a curve as shown in Fig.2.36 between $\frac{W_i}{f}$ and f for different value of f. It is necessary to satisfy the condition that $\frac{V}{f}$ remains constant on varying f. This may be achieved by running an alternator coupled to DC shunt motor. The speed of the motor may be controlled by armature and field control methods to give desired frequency. The voltage of alternator is controlled by adjusting the field current of the alternator. Thus, the ratio $\frac{V}{f}$ can be maintained constant.

The generated voltage of alternator is applied to the transformer and readings are taken as shown in table 2.1 for different values of f.

Equation 9 is similar to equation of a line represented by y = mx + c

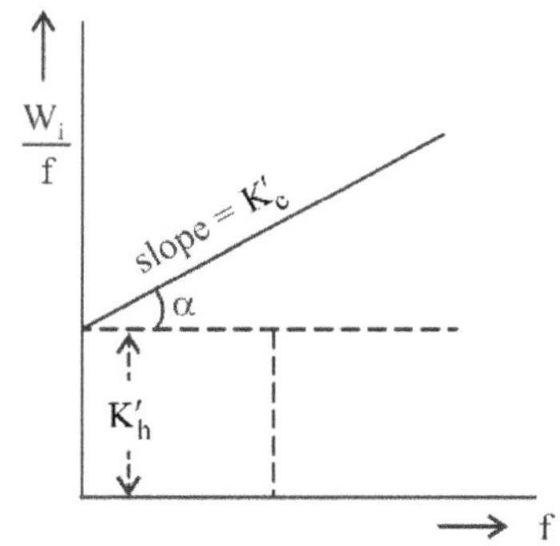

Figure 2.36 Separation of Iron Loss

Here $$y = \frac{W_i}{f} \qquad \text{..... (10)}$$

$$c = K'_h \qquad \text{..... (11)}$$

$$m = \tan\alpha = K'_e \qquad \text{..... (12)}$$

From Fig.2.36 and equations 12 and 13, K'_h and K'_e can be determined and from equation 5 we get

Hysteresis loss $$W_h = K'_h f \qquad \text{..... (13)}$$

Eddy current loss $$W_e = K'_e f^2 \qquad \text{..... (14)}$$

Example 2.8: The total iron loss of a transformer was found to be 2000 W at 300 V supply at 50 Hz. When applied voltage was changed to 150 V at 25 Hz, total iron loss was 750 W. Determine Hysteresis and eddy current losses at 300 V at 50 Hz.

Solution:

Equation 9 gives

$$\frac{W_i}{f} = K'_h + K'_e f \qquad \text{..... (1)}$$

At 50 Hz,

$$\frac{2000}{50} = K'_h + K'_e \cdot 50 \qquad \text{..... (2)}$$

At 25 Hz,

$$\frac{750}{25} = K'_h + K'_e \cdot 25 \qquad \text{..... (3)}$$

Solving equation 2 and 3

$$K'_h = 20 \text{ and } K'_h = 0.4$$

∴ Hysteresis loss $W_h = 20 \times 50 =$ **1000 W**

Eddy current loss $W_e = 0.4 \times 50^2 =$ **1000 W**

REVIEW QUESTIONS

1. What is a transformer? Explain its principle of working.
2. Explain briefly the construction of a simple transformer. Why is the transformer core laminated? Why are the lamination joints staggered?
3. Give advantage of using cruciform core and stepped core instead of square core.
4. What are the two different types of transformer construction? Distinguish between core type and shall type transformers.
5. What are mutual flux and leakage flux? How is leakage flux kept low?
6. What is an ideal transformer and how does it differ from an actual transformer?
7. Explain the function of no-load current of a transformer. What are its two components? Draw a phasor diagram under no load condition.
8. Derive EMF equation of a transformer and show that the induced e.m.f lags the flux producing the emf by $90°$.
9. Develop exact and approximate equivalent circuits of a transformer. Explain what errors get introduced when we change the exact equivalent circuit to approximate equivalent circuit.
10. What is meant by equivalent resistance and leakage reactance referred to primary and secondary sides? Derive expressions for total equivalent resistance and leakage reactance referred to the secondary side.
11. Define Regulation of a transformer. Obtain exact and approximate expressions for determining regulation.
12. Define per unit quantities. Give expression of regulation in pu system.
13. At what power factor, is the regulation zero? Derive the pf for which maximum regulation occurs and there from determine the maximum value of regulation.
14. What are the different types of losses in a transformer? Distinguish between constant losses and variable losses. If copper loss on full load is P_{eu} (W). What would be the copper loss at half and quarter full load.
15. Define efficiency of a transformer. Obtain condition for maximum efficiency. Plot a curve between efficiency and output current and interpret this curve.
16. What is meant by All Day efficiency and what is its practical importance?

17. Explain with circuit diagram the open circuit and short circuit tests on a transformer. Which side of hv/*l*v winding is preferred for OC and SC tests and why?

18. Describe Sumpner's back to back test. Explain how we are able to fully load both the transformers while supplying energy to meet the losses only. Why is this method suitable to measure the temperature rise of large transformers?

19. What is a polarity test and what is its practical utility?

20. Give an experimental method of separating hysteresis and eddy current losses.

PROBLEMS

1. A single phase 2300/230 V, 50 Hz, core type transformer has gross core section of 0.05 m^2. Permissible maximum flux density is 1.1 T. Determine the number of turns on HV and LV windings if the stacking factor be 0.9. [210, 21]

2. A 10 kVA, 400/200 V, single-phase transformers has a primary winding resistance of 0.1 Ω and leakage reactance 0.4 Ω. The corresponding values on the secondary sides are 0.025 Ω and 0.1 Ω respectively. Calculate the voltage required to be applied across the primary winding so as to make the full load current to circulate in the secondary on being short circuited. [20.6 V]

3. A 10 kVA, 2000/400 V single phase transformer has primary winding resistance 5 Ω and leakage reactance 12 Ω. The secondary winding resistance is 0.2 Ω and leakage reactance 0.48 Ω. Determine the secondary terminal voltage and regulation on full load 0.8 pf lagging when the primary applied voltage is 2000 V. [387.6 V, 3.1%]

4. A 50 kVA, 50 Hz, 1-phase transformer has turns ratio of 5:1. The secondary full load current is 200 A. The primary winding and secondary winding resistances are 0.55 Ω and 0.023 Ω respectively. If the transformer is designed for maximum efficiency at two third full load, determine the efficiency when delivering full load at 0.8 power factor lagging. [93, 4%]

5. A 400 kVA distribution transformer has full load copper loss 4 kW and iron loss 3 kW. Determine All Day efficiency if loading cycle in 24 hours is as under.

No. of Hours	Load in kVA	Power factor (lag)
4	300	0.9
8	400	0.8
6	200	0.7
6	0	0

[97.8%]

6. A 5 kVA, 500/250 V, 50 Hz, 1-phase transformer gave the following readings on HV side.

   O.C. test: 500 V, 1 A, 50 W

   S.C. test: 15 V, 6 A, 21.6 W

   Determine (a) Full load copper loss, (b) Efficiency on full load at 0.8 pf lagging, (c) Secondary terminal voltage under this condition, (d) Regulation.

   [(a) 60 W, (b) 97.3%, (c) 240.35 V and (d) 3.86%]

7. Two similar 40 kVA, single – phase transformers gave the following results when tested by Sumpner's back to back method.

   W_1 in supply line (O.C. side) = 800 W

   W_2 in secondaries series circuit at rated current (S.C. side) = 1000 W

   Determine full load iron loss, copper loss and efficiency on full load at unity pf.

   [400 W, 500 W, 97.8%]

CHAPTER 3

THREE PHASE TRANSFORMER

3.1 Construction

Similar to single-phase transformers, the 3-phase transformers have following two types of construction.

1. Three phase core type
2. Three phase shell type

3.2 Three Phase Core Type Transformer

A 3-phase core type transformer with *lv* and *hv* windings is shown in Fig.3.1. The core consists of three vertical legs with two horizontal yokes for completing the magnetic flux path. Each vertical limb carries one phase of primary and secondary windings. First low voltage (*lv*) winding is placed next to the core and *hv* winding is placed over the *lv* windings with proper insulation between core and *lv* winding as well as between *lv* and *hv* windings.

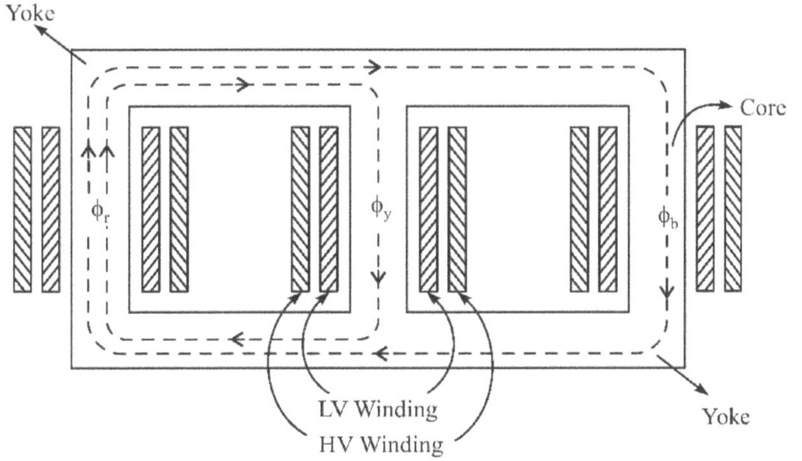

Figure 3.1 Three-phase core type transformer

The sectional areas of three vertical limbs and horizontal yokes are equal. It is possible to build core with only 3 limbs as the fluxes established by three windings are 120° apart in time phase. The flux developed in one limb completes the returns path through other two limbs. This happens in succession in all the limbs sinusoidally.

For constructing the transformer core, thin Si-steel laminations of 0.14, 0.28, 0.3 mm or 0.35 mm thickness are used. The silicon content is 1-4% and the rest part is iron. The laminations have an insulation coating of enamel. As a result of this the effective iron section is reduced by 9-12 % of the gross iron section.

When balanced three phase sinusoidal voltages are applied to the primary windings, the fluxes ϕ_r, ϕ_y and ϕ_b produced in the limbs shall also be sinusoidal in nature having a phase displacement of 120°. These fluxes shall be lagging the supply voltages by 90° and induce emfs in windings the windings lagging the respective fluxes by 90°. The induced emfs in the windings lag the supply voltages by 180°. Thus, the induced emf would be opposing the supply voltages in conformity with the Lenz's law. The flux path produced by the middle leg shall be somewhat shorter than that produced by the outer legs. This will make a little imbalance in the magnetizing currents of the three legs, smaller magnetizing current being for the middle leg. This imbalance is very little and is ignored in practice.

The 3-phase core type transformer has only two windows and each window contains two *lv* and two *hv* windings as shown in Fig.3.1.

Sometimes 5 limb transformers are also used. They have two outer limbs of half the core section as shown in Fig.3.2. In 5-limb transformer, the windings are placed on the central three limbs as in the case of 3-limb core type transformer. As shown in Fig.3.2 the fluxes of three phase are independent and Sectional area of upper and lower Yokes, and also of the outer Yokes is half of the core section of the three central limbs. The height of this transformer is less as compared to 3 phase 3 limb transformer. It may be noted that in 3 limb core type transformer, two core limbs act as return path for the flux in the third limb. Flux flows upwards in corresponding limb and downwards in the other two limbs. The magnetic circuits for different phases are in series and are interdependent.

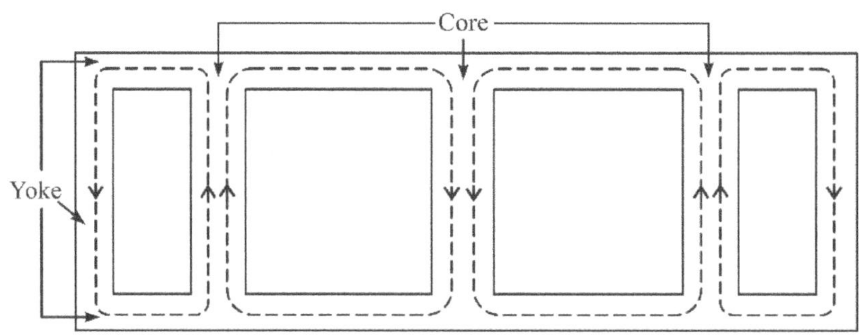

Figure 3.2 Five limb transformer having three middle limbs as core

3.3 Three-Phase Shell Type Transformer

The stamping (lamination) for a 3-phase shell type transformer is shown in Fig.3.3. It would appear as three single-phase shell type transformer have been stacked together to make a 3-phase shell type transformer. The vertical three middle legs are called core and the rest part for completing the flux path is termed yoke. The sectional area of outer yoke is half that of middle core section.

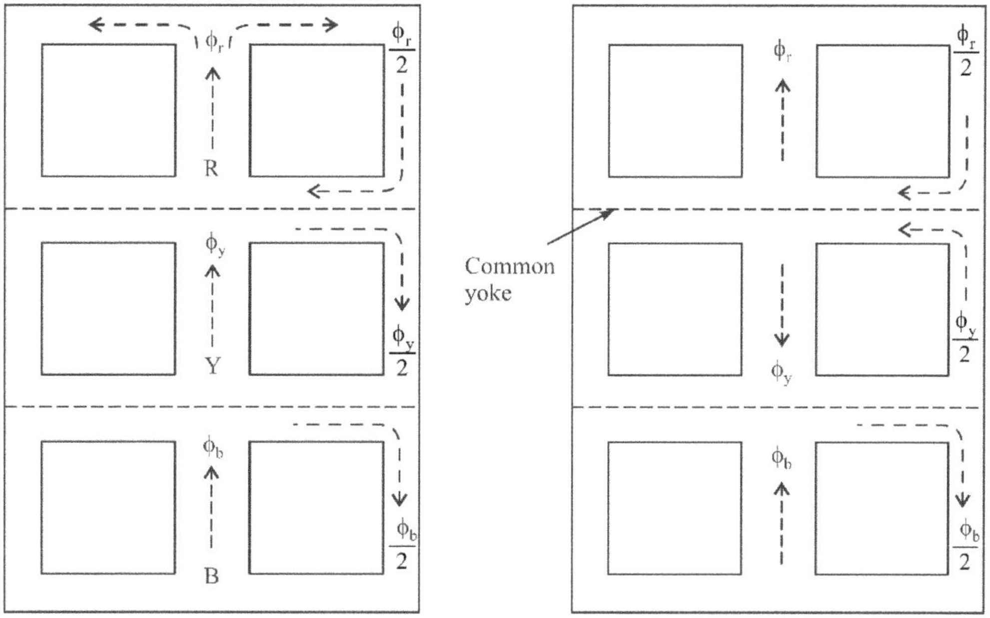

(a) Fluxes in common yoke when applied voltages are in regular sequence R-Y-B.

(b) Fluxes in common yoke when applied voltage is reversed in middle winding.

Figure 3.3 Fluxes in common yoke in 3 phase shell type transformer

When the 3-phase applied voltages are in regular sequence R-Y-B the fluxes produced in three core parts shall be ϕ_r, ϕ_y, and ϕ_b, all in upward directions. In common yoke the vector difference of two fluxes $\phi_r/2$ and $\phi_y/2$ as shown in diagram of Fig.3.3(a) will be produced.

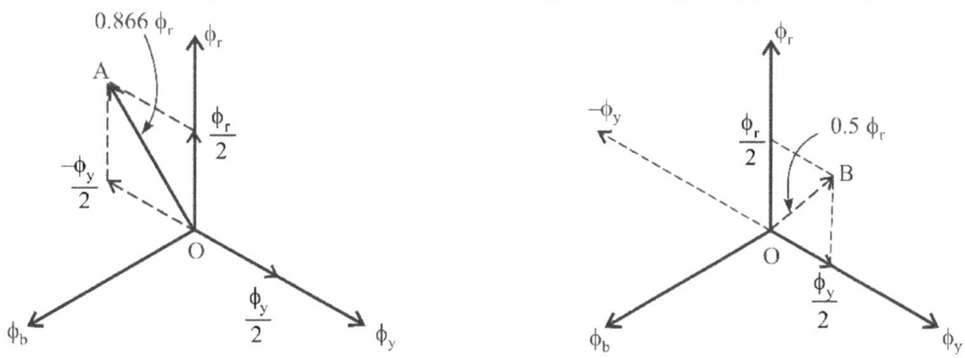

(a) (b)

Figure 3.4 Phase diagrams to show the fluxes in common yoke

In Fig.3.4(a) the RMS values of the fluxes ϕ_r, ϕ_y, ϕ_b are shown 120° apart from each other. These fluxes will divide into two parts $\phi_r/2$, $\phi_y/2$ and $\phi_b/2$. In the common yoke between phase R and Y, the flux shall be vector difference of the flux $\phi_r/2$ and $\phi_y/2$. The flux $\phi_y/2$ is plotted as $-\phi_y/2$ in Fig. 3.4(a). Thus, the resultant flux is common yoke will be

OA = Vector sum of $\phi_r/2 + (-\phi_y/2)$

$$= \frac{\sqrt{3}}{2}\,\phi_r = 0.866\,\phi_r$$

Fig.3.4(b) shows the phasor diagram, when the voltage in the middle winding Y has been reversed. Now the flux in common yoke between phases R and Y will be the vector sum of fluxes $\phi_r/2$ and $\phi_y/2$ and shall be given by the vector.

OB = Vector sum of fluxes $\phi_r/2 + \phi_y/2$

$= \phi_r/2 = 0.5\,\phi_r$.

Thus, the reversing the supply in the middle winding the flux in common yoke reduces from $0.866\,\phi_r$ to $0.5\,\phi_r$ which results in saving of iron in common yoke.

3.4 Three-Phase Unit Transformer versus Bank of 3 Single-Phase Transformers

For three phase transformers two types of arrangements are possible. Either to use a single unit Three-phase transformer which may be Core type Star or Shell type, or use 3 single phase transformers and connect them externally to form Star or Delta connection. Three single-phase transformers shall again be single-phase core type or shell type. Both arrangements have merits as well demerits.

(a) A 3-phase core type transformer unit costs about 15% less than a Bank of three 1-phase core type transformers and also occupies less space.

(b) A 3-phase core type unit is slightly more efficient than a Bank of three-1-phase transformers.

(c) The expensive high voltage terminals (bushings) are reduced to three instead of six for three 1-phase transformers.

(d) The switchgear, busbar arrangement and wiring of a single unit 3-phase transformer are somewhat less costly and simpler than those for a Bank of transformers.

(e) From transport consideration three 1-phase transformers are better.

(f) The bank offers the advantage of a derated (58%) power on open delta operation, when one 1-phase transformer becomes in-operative. whereas if one phase winding of single 3-phase transformer breaks, it has to be removed from the service interrupting the supply.

(g) One 1-phase transformer in a bank may be provided with higher/lower rating than the others in order to supply an unbalanced load.

(h) Only one spare 1-phase transformer is needed when bank of 1-phase transformer is used. Whereas one complete 3-phase transformer unit will be required to keep as spare.

(i) In the 3-phase bank, the phases are electrically connected but the 3 magnetic circuits are independent. Whereas in 3 phase 3 limb core type transformer, the three magnetic circuits are also linked. For de-linking the magnetic circuits 5-limb transformer is used.

(j) In 3-phase, shell type unit transformer same material is required for bank of three single-phase shell type transformer. The magnetic circuits remain separate.

3.5 Three-Phase Transformer Connections

A three-phase transformer may have either Star or Delta connections. We may consider here bank of three 1-phase transformers to be connected in either of the following 4 ways.

1. Delta-Star connection
2. Star-Delta connection
3. Star-Star connection
4. Delta-Delta connection

3.5.1 Delta-Star Connection

This type of connection is very useful at sending end and receiving end sub-stations in power systems. At power stations the generator feeds the delta winding and the star winding is connected to *hv* transmission lines. This scheme of connection is useful where stepping up of the voltage is required. In distribution system, feeders are connected to delta winding and the star-winding provides 3-phase, 4-wire supply to consumers. The delta-star connection of transformers and the phasor diagram are shown in Fig.3.5 and Fig.3.6 respectively.

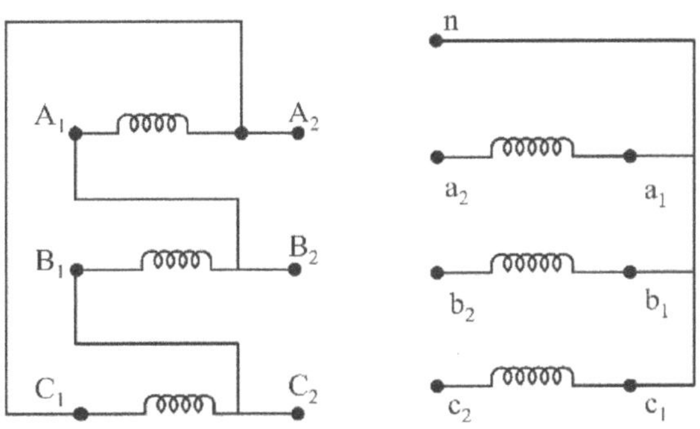

Figure 3.5 Delta-Star connection

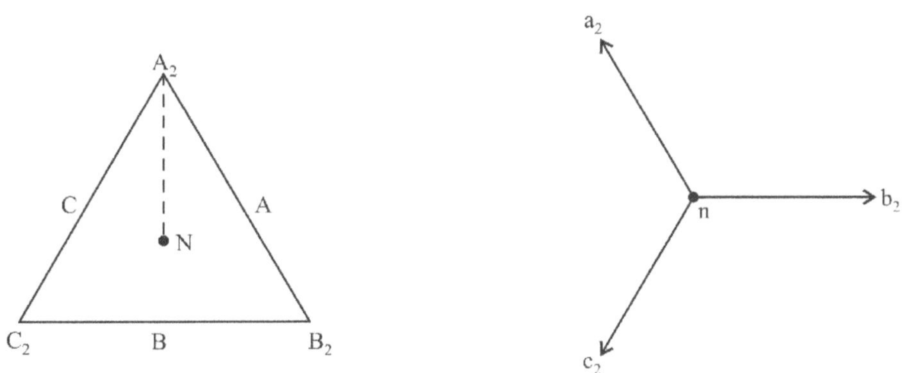

Figure 3.6 Phasor diagram of primary and secondary voltages

On the primary delta side, the line and phase voltages are same and on the secondary side the line voltage is $\sqrt{3}$ times the phase voltage. There appears a phase displacement of 30° between the primary and secondary voltages. In delta-star connection, the presence of delta winding helps in stabilizing potential of star point thus solving the problem of third harmonic component. By reversing the connection on either side of the winding, the secondary system voltage can be made to displace by 30° in the opposite direction.

3.5.2 Star-Delta Connection

The connections and phasor diagram for Star-Delta connection are shown in Fig.3.7 and 3.8 respectively. Here on primary side line voltage is $\sqrt{3}$ times the phase voltages while on secondary side line and phase voltages are equal.

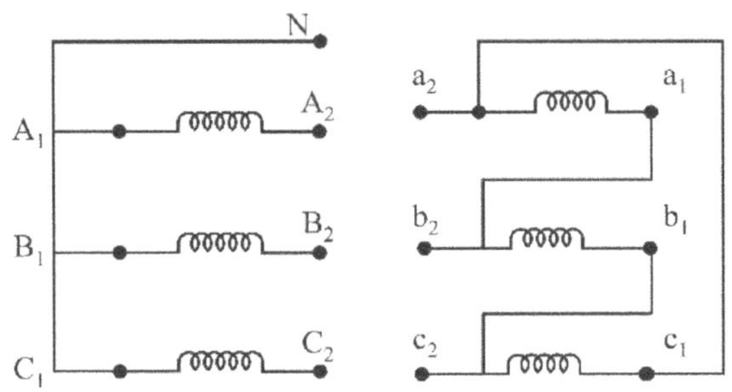

Figure 3.7 Star-Delta connection

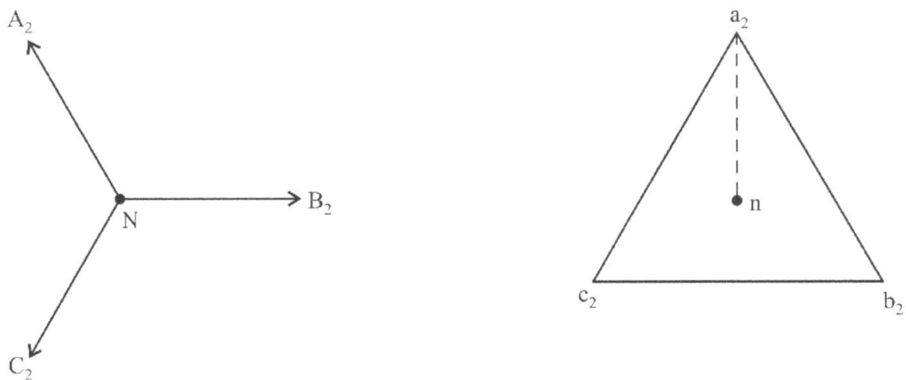

Figure 3.8 Phasor diagram of primary and secondary voltages

Because of the delta winding the third harmonic components are consumed in circulating third harmonic current in the closed delta without requirement of neutral wire. Thus, there is no problem with unbalanced loads and third harmonics.

In this type of connection there is a phase shift of $\pm$ 30° between the primary and secondary voltage, which has to be taken in account carefully while operating them in parallel.

It may be seen that star-delta connection can be obtainable by interchanging the role of primary and secondary in delta-star connection.

3.5.3 Star-Star Connection

The Star-Star connection of the transformers and the phasor diagram of voltage are shown in Fig.3.9 and 3.10 respectively.

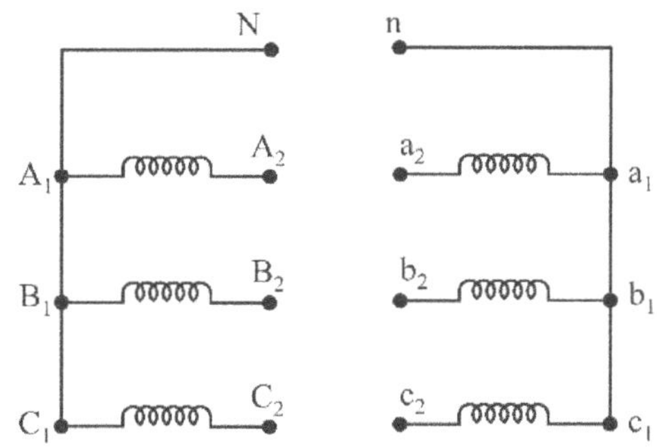

Figure 3.9 Star-Star connection

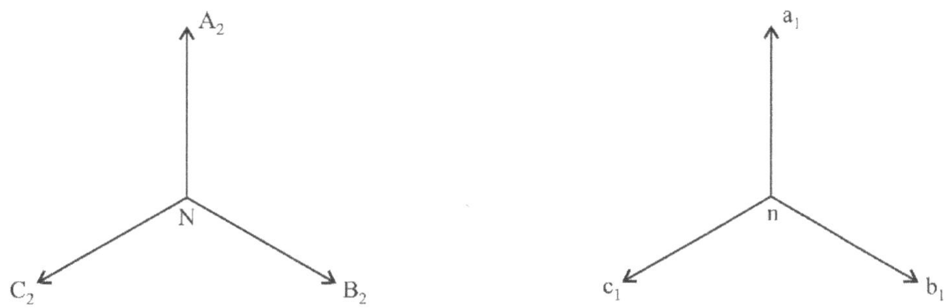

Figure 3.10 Phasor diagram of primary and secondary voltages

This connection does not cause any phase displacement between primary and secondary. If a phase displacement of 180° is desired, the supply can be given to terminals a_1, b_1 and c_1 and a_2, b_2 and c_2 can be connected together to form the neutral. The performance is same for both connections. When two Star-Star connected transformers are to be put in parallel, both of them must be connected either for zero phase displacement or 180° phase displacement.

There are two serious problems with star-star connection.

1. If there is no neutral and the load is unbalanced the phase voltages tend to become highly unbalanced. Hence this connection is not suitable for unbalanced loads unless there is neutral connections.
2. The magnetizing current of any transformer is sufficiently non-sinusoidal and contains large third harmonic component. The third harmonic magnetizing current in three phase add together and need a path to flow through the neutral.

   Both for unbalance and third harmonic problem of the star-star connection, one of the following methods can be applied.

   (a) Solid grounding of neutral. It provides a path for third harmonic currents to flow besides providing a return path for unbalanced currents of unbalanced loads.

   (b) Providing tertiary windings. Where Star-Star connection is without neutral, a third winding, in addition to primary and secondary windings, is provided. This is connected in closed delta.

3.5.4 Delta-Delta Connection

The Delta-Delta connection of three phase transformers and the phasor diagram are given in Fig.3.11 and 3.12 respectively. The line voltage is equal to the phase voltage and the line current is $\sqrt{3}$ times the phase current.

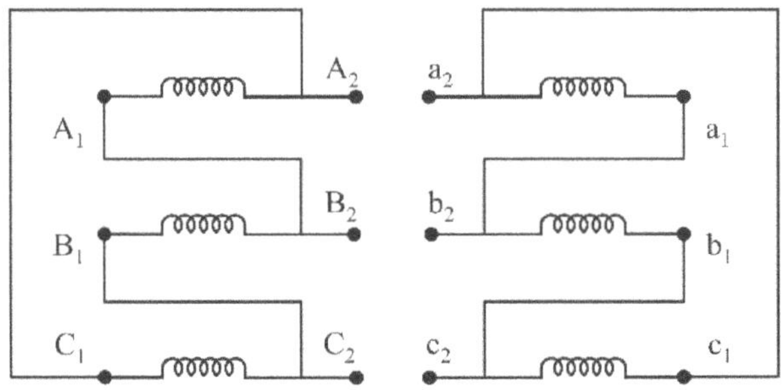

Figure 3.11 Delta-Delta connection

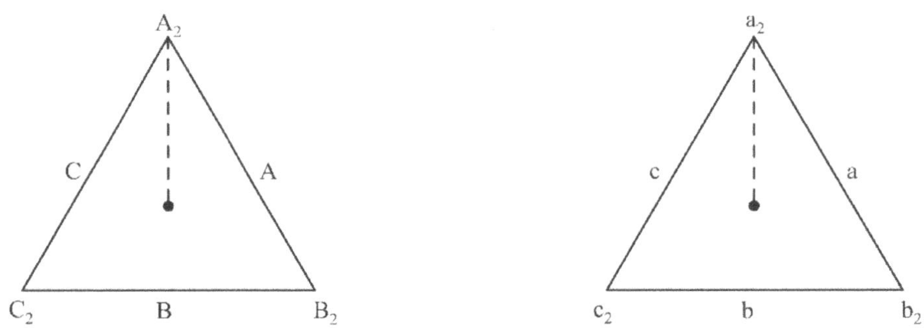

Figure 3.12 Phasor diagram of primary and secondary voltages

There is zero or 180° phase displacement depending on the internal connections of the transformers. The closed delta provides a path for the third harmonic currents and clamps out the third harmonic voltages.

A Delta-Delta connected transformer has no problem with unbalanced loads and harmonics. There is also no phase shift associated with it.

3.6 Three Phase Transformer Phasor Groups (Vector Groups)

There are four phasor groups and each group has two methods of connections of *hv* and *lv* windings, which give the same phase displacement between the windings. All connections which lead to same phase shift are called Vector Groups.

Phase displacement is expressed as the clock hour number and is designated 0, 6, 1 and 11 O' clock of the clock. As an example, a transformer with vector group symbol Yy0 will mean a transformer having both primary and secondary windings as 'Star' connected and belonging to groups with zero degree phase displacement. The method of labeling of transformer terminals is given in Fig.3.13. Terminals A_1 and A_2 refer to *hv* side and a_1 and a_2 to *lv* side. A_2 and a_2 have +ve polarity with respect to A_1 and a_1 having –ve polarity. The phasor diagram of voltage is given in Fig.3.14.

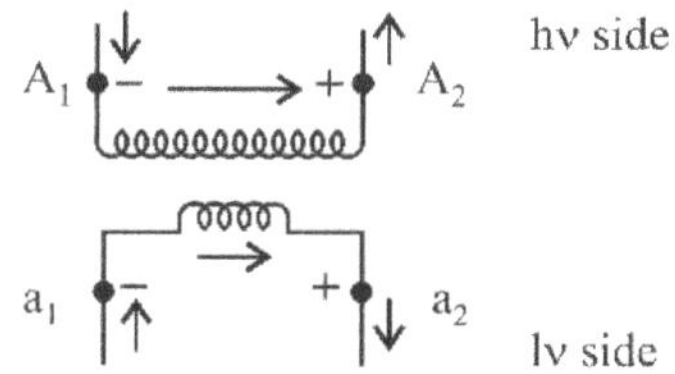

Figure 3.13 Labelling of transformer terminals

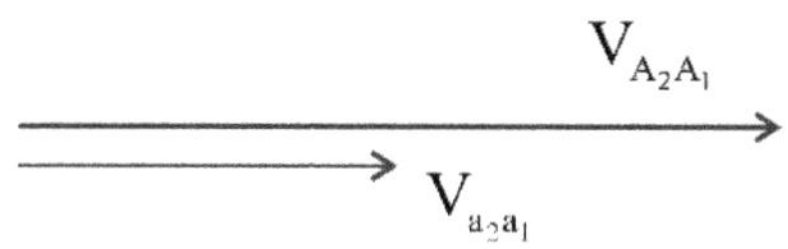

Figure 3.14 Phasor diagram with zero phase displacement

The vector symbols of different phasor groups are given in Table 3.1.

Table 3.1 Vector Symbols

Group Number	Winding Connection Primary/ Secondary	Phase Displacement	Clock hour number	Vector Symbol
1.	⅄ / ⅄	0°	0	Yy0
	Δ / Δ	0°	0	Dd0
2.	⅄ / Y	180°	6	Yy6
	Δ / Δ	180°	6	Dd6
3.	Δ / Y	-30°	1	Dy1
	⅄ / ▷	-30°	1	Yd1
4.	Δ / ⅄	$+30^{\circ}$	11	Dy11
	⅄ / ▷	$+30^{\circ}$	11	Yd11

The description of above vector symbols is given below.

3.6.1 Zero Phase Displacement Yy0 Connection

Transformer connections and phasor diagram for zero phase displacement of stat-star connection are shown in Fig.3.15 and Fig.3.16 respectively.

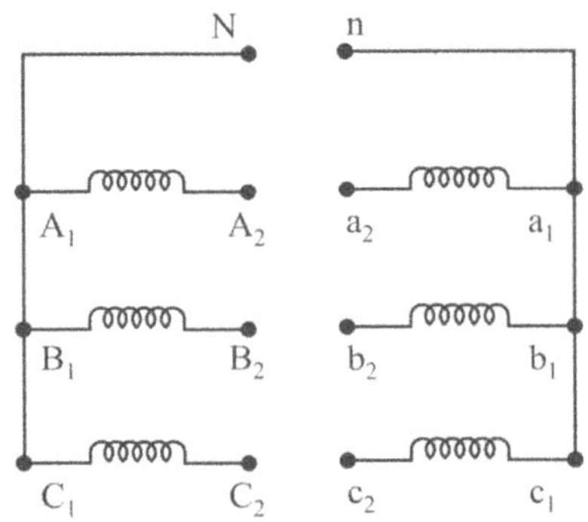

Figure 3.15 Yy0 connection

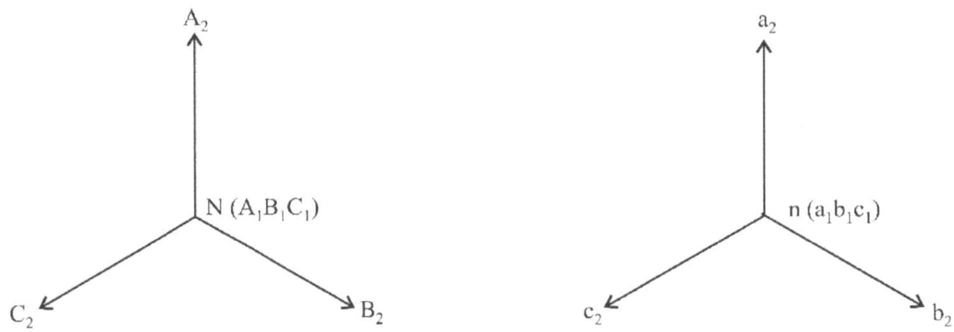

Figure 3.16 Phasor diagram

3.6.2 Zero Phase Displacement Dd0 Connection

The transformer connections and phasor diagram for zero phase displacement of Delta-Delta connection are shown in Fig.3.17 and Fig.3.18 respectively.

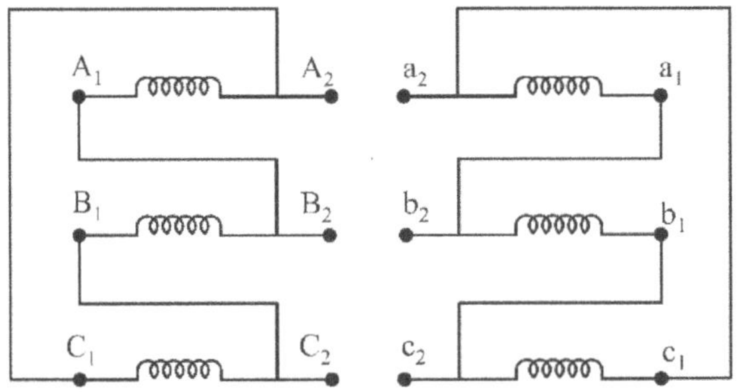

Figure 3.17 Dd0 connection

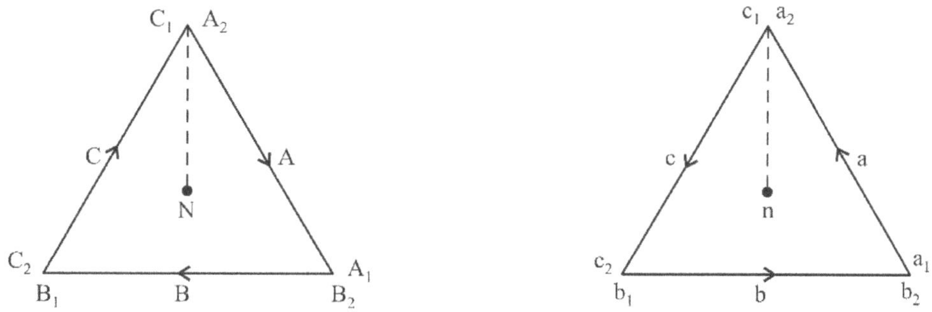

Figure 3.18 Phasor diagram

- The *lv* side phasors are 0° displaced.
- The supply terminals to feed load will be a_2, b_2 and c_2 on *lv* side.

3.6.3 180° Phase Displacement Yy6 Connection

The transformer connection and phasor diagram for 180° phase displacement are shown in Fig.3.19 and Fig.3.20 respectively.

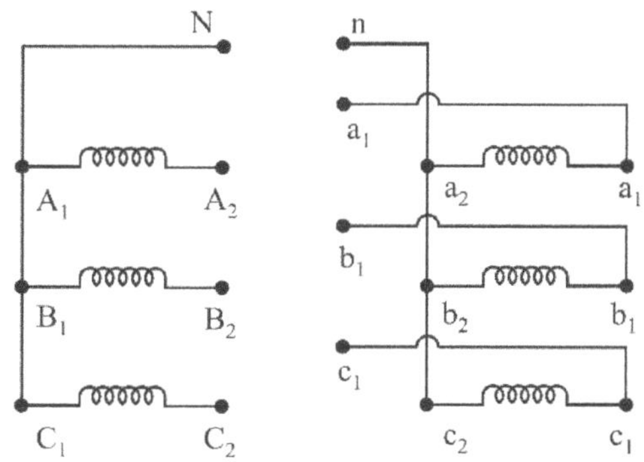

Figure 3.19 Yy6 connection

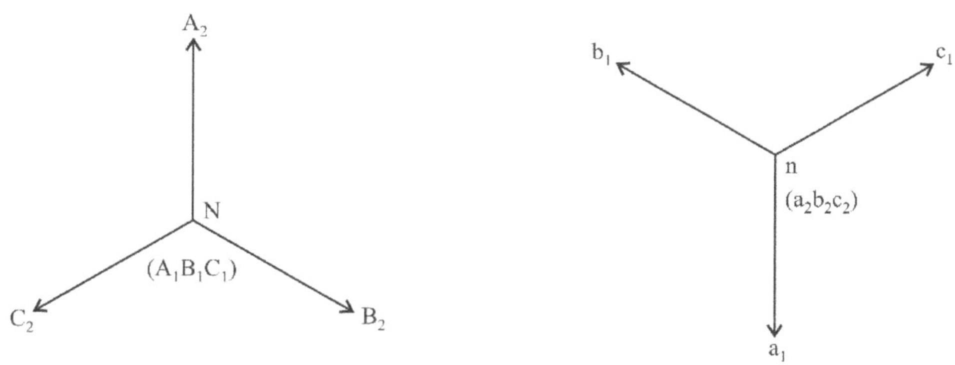

Figure 3.20 Phasor diagram

- 180° phase displacement is achieved by reversing the connection on the *lv* side.
- The supply terminals to feed load will become a_1, b_1 and c_1 and neutral would be n (a_2 b_2 c_2).

3.6.4 180° Phase Displacement Dd6 Connection

The connections and phasor diagram for 180° phase displacement are shown in Fig.3.21 and Fig.3.22 respectively.

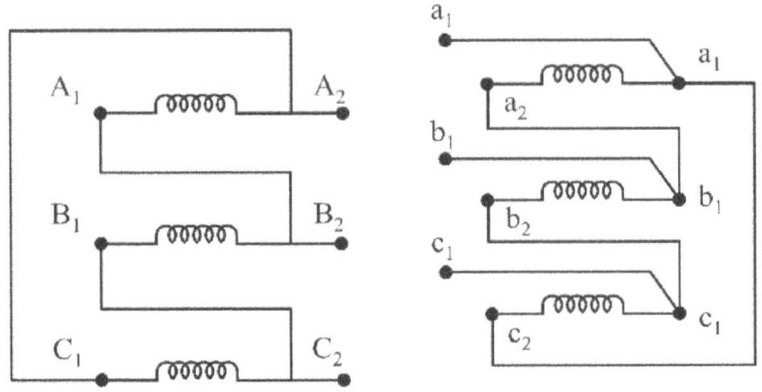

Figure 3.21 Dd6 connection

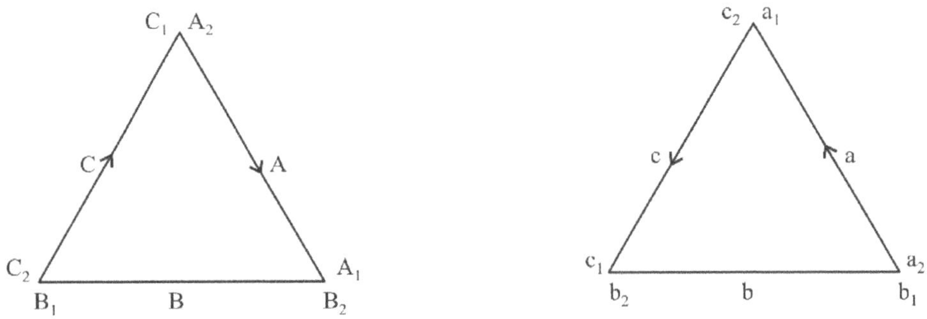

Figure 3.22 Phasor diagram

- The *lv* side phasor are 180° phase displaced.
- The supply terminals to feed load will be a_1, b_1 and c_1 on *lv* side

3.6.5 -30° Phase Displacement Dy1 Connection

The connection and phasor diagram for –30° phase displacement are shown in Fig.3.23 and Fig.3.24 respectively which are obtained by the following procedure.

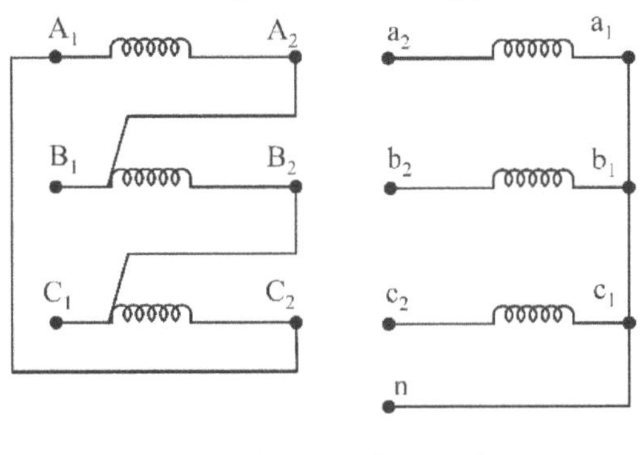

Figure 3.23 Dy1 Connection

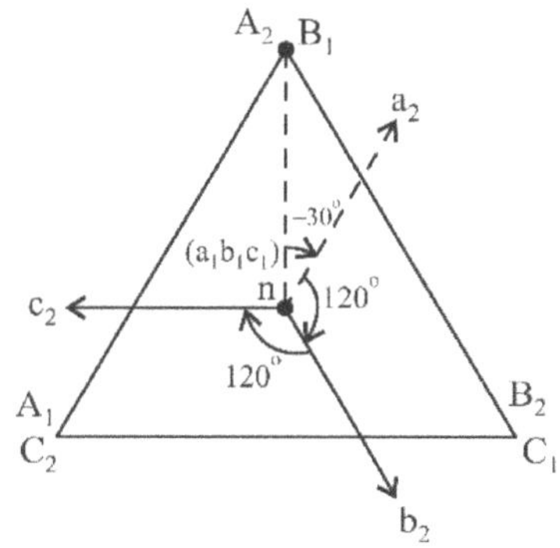

Figure 3.24 Phasor diagram

- Draw a vertical line n A_2
- Draw a line na_2 making an angle of –30° (i.e., clock hour 1).
- Draw a line nb_2 at 120° and nc_2 another 120° lagging behind na_2.
- The neutral point *n* of *lv* winding is termed a_1 b_1 c_1.
- From point A_2 draw A_2 A_1 parallel to a_2 a_1 and B_1 B_2 parallel to b_1 b_2.
- Draw a line C_2 C_1 parallel to c_2 c_1.
- Thus, the connections on Delta side are obtained as A_2 B_1, B_2 C_1 and C_2 A_1.
- With these connections a phase difference of –30° is introduced between primary and secondary voltages.

3.6.6 -30° Phase Displacement Yd1 Connection

The connections and phasor diagram for –30° phase displacement are shown in Fig.3.25 and Fig.3.26, which are obtained by the following procedure.

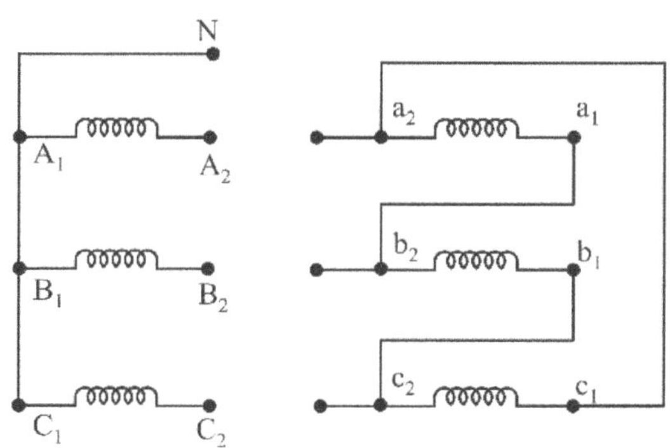

Figure 3.25 Yd1 Connection

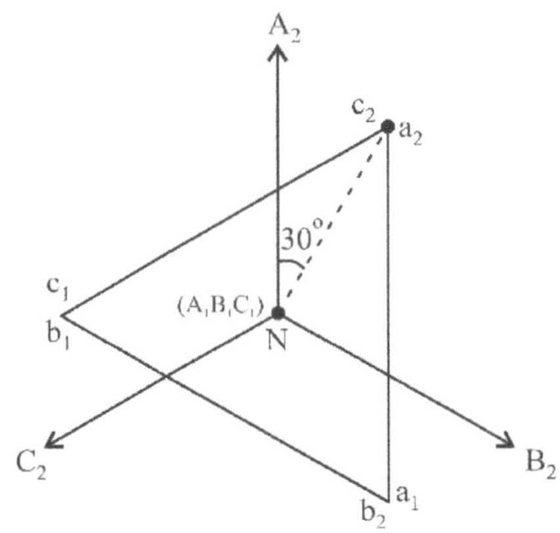

Figure 3.26 Phasor diagram

- Draw the star vector A_2 B_2 C_2 with N as neutral point which is the junction is of A_1 B_1 C_1.
- From point N draw a line Na_2 at –30° and from point a_2 draw lines a_2 a_1 parallel to A_2 A_1.
- From point a_1 draw a line b_2 b_1 parallel to B_2 B_1.
- From point a_2 draw a line c_1 c_2 parallel to C_1 C_2.
- Thus, the connections on Delta side are obtained as a_1 b_2, b_1 c_2 and c_1 a_2.
- With these connections a phase displacement of -30° is obtained between the primary and secondary sides.

3.6.7 30° Phase Displacement Dy11 Connection

The connections and phasor diagrams for +30° phase displacement are shown in Fig.3.27 and Fig.3.28, which are obtained by the following procedure.

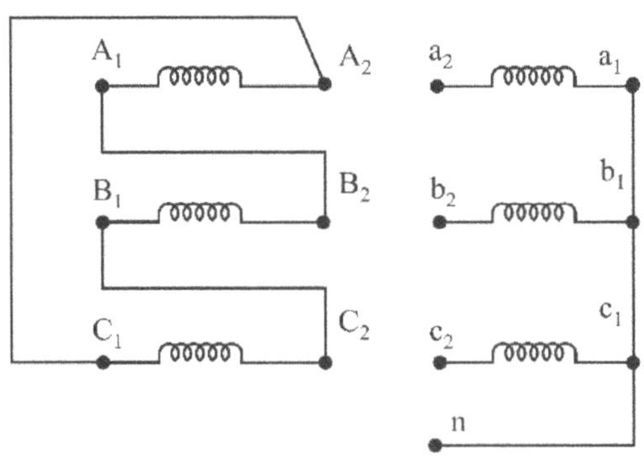

Figure 3.27 Dy11 Connection

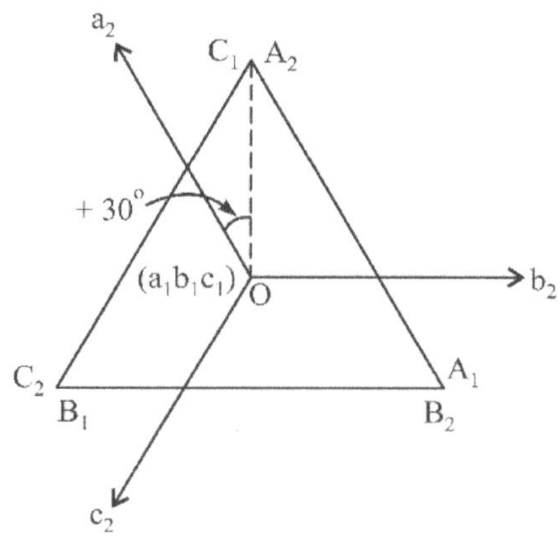

Figure 3.28 Phasor diagram

- Draw vertical line nA_2 and draw na_2 making an angle of +30° from nA_2.
- Draw nb_2 and nc_2 making angle of 120° and 240° lagging from na_2. The neutral point n is the junction of a_1 b_1 c_1.
- From point A_2 draw line A_2 A_1 parallel to a_2 a_1 and from point A_1 draw B_2 B_1 parallel to b_2 b_1.
- From point C_2 draw line C_1 C_2 parallel to c_1 c_2.
- Thus the connections on delta side are A_1 B_2, B_1 C_2 and C_1 A_2.

3.6.8 30° Phase Displacement Yd11 Connection

The connections and phasor diagram for +30° phase displacement are shown in Fig.3.29 and Fig.3.30. which are obtained by the following procedure.

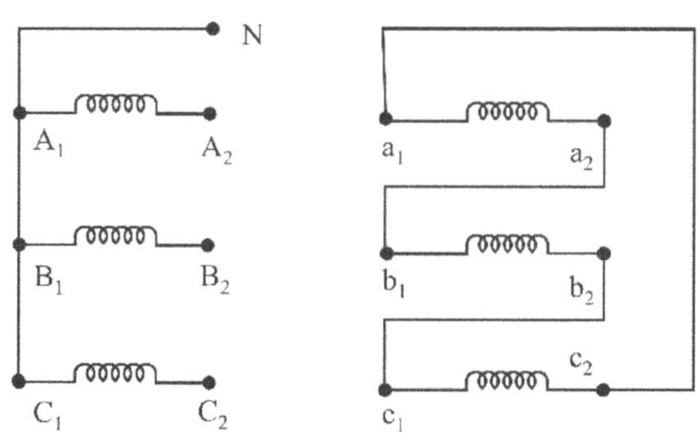

Figure 3.29 Yd11 Connection

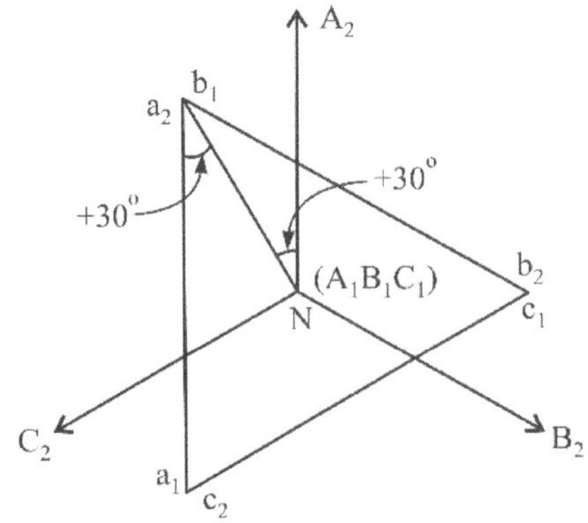

Figure 3.30 Phasor diagram

- For star side draw phasor NA_2 vertical and draw NB_2 and NC_2 displaced at 120° and 240° behind NA_2.
- Draw vector Na_2 at +30° displacement from NA_2.
- From point a_2 draw line $a_2\ a_1$ at 30° from NA_2 and another line $b_1\ b_2$ again from point a_2 at 30°.
- From point a_1 draw line $c_2\ c_1$ parallel to $C_2\ C_1$ meeting at point b_2.
- Thus, the connections on delta side are $a_2\ b_1$, $b_2\ c_1$ and $c_2\ a_1$.

3.7 Effect of Connections on Parallel Operation of Transformers

- Three phase transformers having same phasor group number can be operated satisfactorily in parallel [Refer to Table 3.1].
- Three phase transformers belonging to phasor group 3 can be operated in parallel with transformers of phasor group 4 by interchanging any two external connections to the supply and also the corresponding two external *lv* connections to the load bus bar. There is no need to change any internal connection.
- Transformers belonging to phasor group 1 and 2 cannot be operated in parallel unless the internal connections of one of the transformers are suitably changed.

3.8 Open Delta (V-V) Connection

While working on three single-phase transformers as Bank of transformers with delta connection and if due to any fault one single-phase transformer of the Bank becomes inoperative, yet derated power can be obtained from open delta operation as shown in Fig.3.31 and Fig.3.32.

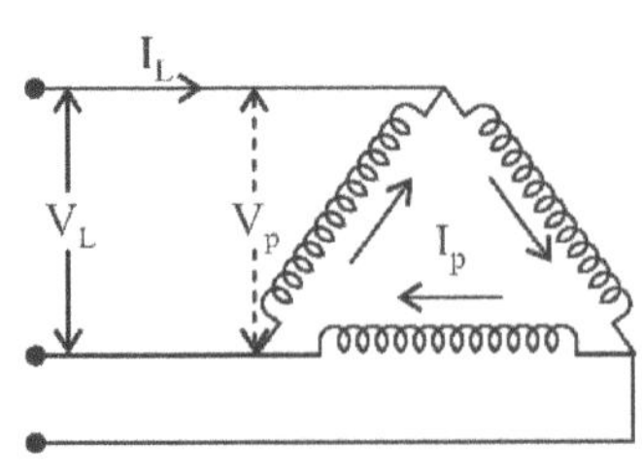

Figure 3.31 Delta-connection

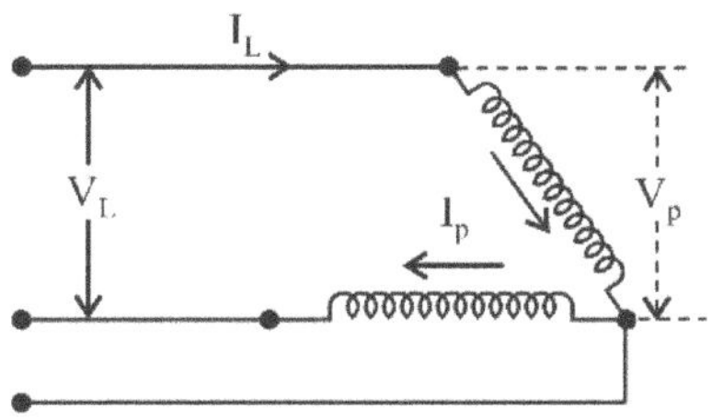

Figure 3.32 Open-delta or V-connection

Let V_L = Line voltage

V_P = Phase voltage = V_L

I_L = Line current

$= \sqrt{3}\, I_p$

I_P = Phase current

For healthy delta-connection Fig.3.31.

$$\text{Power} \quad P = \sqrt{3}\ V_L\, I_L$$

$$= \sqrt{3}\ .\ V_P\ .\ \sqrt{3}\, I_p$$

$$= 3\ V_P\, I_p$$

With open delta connection Fig.3.32

$I_L = I_p$

Power with open delta is

$$\text{Power} \quad P' = \sqrt{3}\ V_L\, I_L$$

$$= \sqrt{3}\ V_P\, I_P$$

Ratio of power with open delta connection and healthy delta connection

$$= \frac{\text{Power open delta}}{\text{power with healthy delta}} = \frac{P'}{P}$$

$$= \frac{\sqrt{3} V_p I_p}{3 V_p I_p}$$

$$= \frac{1}{\sqrt{3}} = 58.8\%$$

So a derated power of 58.8% of the normal power shall be available.

3.9 Open Star Connection

Working with three single-phase transformers as Bank of transformers having star connection, if, due to any fault, one single-phase transformer of the Bank becomes inoperative; yet derated power can be obtained from open star operation. This is explained in Fig.3.33 and Fig.3.34.

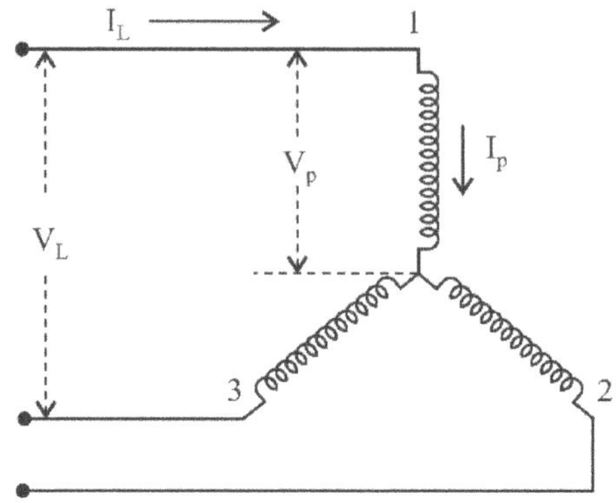

Figure 3.33 Star-connection

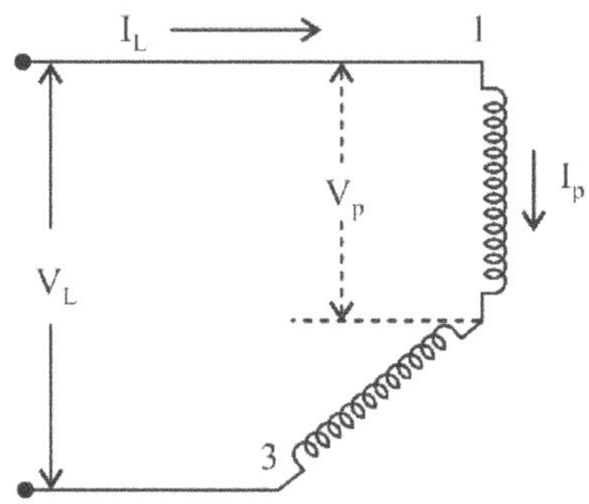

Figure 3.34 Open star connection

Let V_L = Line voltage = $\sqrt{3}\,V_P$

V_P = Phase voltage

I_L = Line current

$= I_p$

For healthy star-connection Fig.3.33

$$\text{Power} \quad P = \sqrt{3}\, V_L I_L$$

$$= \sqrt{3}\,.\,\sqrt{3}\, V_P I_P$$

$$= 3\, V_P I_P$$

with open star Fig.3.34, phase voltage is

$$V_P' = \frac{V_L}{2} = \frac{\sqrt{3}V_p}{2}$$

where V_p = Phase voltage with healthy star

Here again phase current

$I_P = I_L$

Since only 2 phases are operative, so derated available power is

$$P_{derated} = 2 \times V_p' \times I_p$$

$$= 2 \times \frac{\sqrt{3}V_p}{2} \times I_p$$

$$= \sqrt{3}\ V_P I_p$$

Ratio of power with open star connection and healthy star connection

$$= \frac{P_{derated}}{\text{power with healthy star}}$$

$$= \frac{\sqrt{3}V_pI_p}{3V_pI_p}$$

$$= \frac{1}{\sqrt{3}} = 58.8\%$$

So, a derated power of 58.8% of the normal power of healthy star connection shall be available.

3.10 Three Phase to Two Phase (Scott-Connected) Transformation

A scott-connected transformer is used to obtain a balanced 2-phase supply from a balanced three phase supply system and vice versa. This is achieved by using two identical single-phase transformers, each having N1/N2 turns. The secondary of one (Main) transformer has a center tapping, while the secondary of the second (Teaser) transformer has a tapping at $\sqrt{3/2}$, i.e., 86.6%. The center-tap point D, of the secondary winding CB of the main transformer, is connected at point D to the secondary winding of the Teaser transformer as AD having 86.6% turns. The primary windings a-b and b-c are placed connected at 90° phase displacement with each other. The arrangement of the windings is shown in Fig.3.35 and the corresponding phasor diagram is drawn in Fig.3.36.

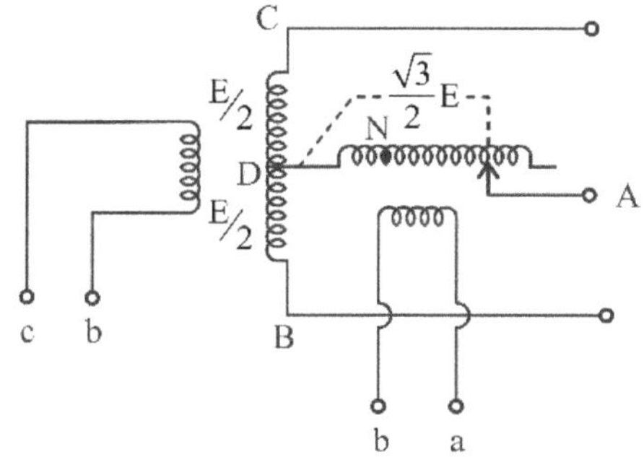

Figure 3.35 Scott-connected transformer

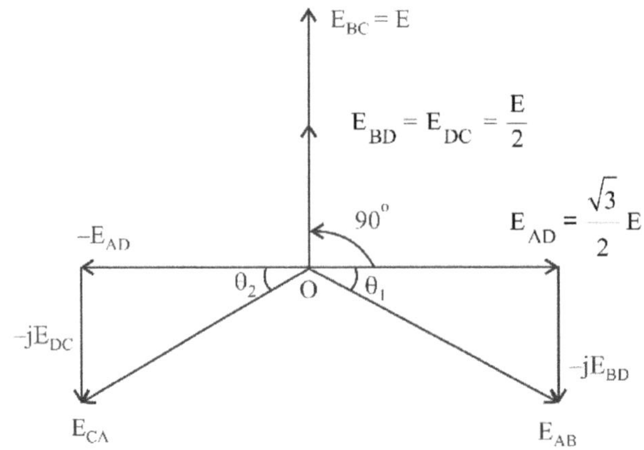

Figure 3.36 Phasor diagram for Scott connected transformer

Consider that a 2 phase supply is given to terminals a-b and b-c. As per ratio of turns in main and teaser transformers, the induced emf across terminals BC shall be E and across terminals AD will be $\frac{\sqrt{3}}{2}E$ and these emf's shall be $90°$ displaced with respect to each other. Draw the phasor diagram Fig.3.36 making $E_{AD} = \frac{\sqrt{3}}{2}E$ in horizontal direction and $-E_{AD}$ in opposite direction to E_{AD}. From points O draw vector $E_{BC} = E$, $90°$ displaced to vector E_{AD}. The midpoint of this vector shall represent $E_{BD} = E_{DC} = \frac{E}{2}$.

The voltage between terminals A and B is given by

$$E_{AB} = E_{AD} + j\,E_{DB}$$

$$= E_{AD} - jE_{BD}$$

$$= \frac{\sqrt{3}}{2}|E| - j\frac{1}{2}|E|$$

$$= |E|\sqrt{\frac{3}{4} + \frac{1}{4}} = |E|$$

Angle θ_1 is given by

$$\theta_1 = \tan^{-1}\frac{E_{BD}}{E_{AD}} = \tan^{-1}\frac{\frac{E}{2}}{\frac{\sqrt{3}}{2}E}$$

$$= \tan^{-1}\frac{1}{\sqrt{3}} = 30°$$

Thus, the angle between voltages E_{AB} and E_{BC}

$$= 30° + 90° = 120°$$

The voltage between terminals B and C is $E_{BC} = |E|$

Further, the voltage between terminals C and A is

$$E_{CA} = -E_{AC} = -(E_{AD} + j\,E_{DC})$$

$$= -E_{AD} - j\,E_{DC}$$

$$= -\frac{\sqrt{3}}{2}|E| - j\left|\frac{E}{2}\right|$$

$$= |E|\sqrt{\frac{3}{4}+\frac{1}{4}}$$

$$= |E|$$

Angle θ_2 is given by

$$\theta_2 = \tan^{-1}\frac{E_{DC}}{E_{AD}} = \tan^{-1}\frac{\frac{E}{2}}{\frac{\sqrt{3}}{2}E}$$

$$= \tan^{-1}\frac{1}{\sqrt{3}} = 30^\circ$$

Thus, the angle between E_{CA} and E_{BC}

$$= 30^\circ + 90^\circ = 120^\circ$$

Further, the voltage between terminate B and C is

$$E_{CA} = -E_{AC} = -(E_{AD} + j\,E_{DC})$$

$$= -E_{AD} - j\,E_{DC}$$

$$= \frac{-\sqrt{3}}{2}|E| - j\left|\frac{E}{2}\right|$$

$$= |E|\sqrt{\frac{3}{4}+\frac{1}{4}}$$

$$= |E|$$

Angle E_2 is given by

$$\theta_2 = \tan^{-1}\frac{E_{DC}}{E_{AD}} = \tan^{-1}\frac{\frac{E}{2}}{\frac{\sqrt{3}}{2}E}$$

$$= \tan^{-1}\frac{1}{\sqrt{3}} = 30^\circ$$

Thus, the angle between E_{CA} and E_{BC}

$$= 30^\circ + 90^\circ = 120^0$$

Hence it is observed that a balanced two phase supply or two 1-phase supply displaced at 90° would give a balanced three phase supply. The 3 phase voltages shall be equal in magnitude and displaced at 120° from each other. We may now locate a neutral point N on the teaser transformer. At the neutral point the voltage

$$E_{AN} = E_{BN} = E_{CN} = \frac{1}{\sqrt{3}} E = 0.577 \text{ E, whereas}$$

$$E_{AD} = \frac{\sqrt{3}}{2} E = 0.866 \text{ E}$$

Voltage between point D and N

$$E_{DN} = E_{AD} - E_{AN}$$

$$= 0.866 \text{ E} - 0.577 \text{ E}$$

$$= 0.289 \text{ E}$$

Voltage between B-N and C-N will be

$$E_{BN} = E\sqrt{0.5^2 + 289^2} = E_{CN}$$

$$= 0.577 \text{ E}$$

In case it is desired to have a neutral point N on the teaser transformer, another tapping point N should be taken out at 57.7% of turns from point A or at 28.9% of turns from point D.

Example 3.1: Two electric furnaces are supplied with single phase supply at 80 V from a 3-phase 11 kV supply by means of two single-phase Scott – connected transformers. The load on one transformer is 500 kW and on the other is 800 kW. What current would flow in each of the 3 phase lines when the furnaces work at

(a) unity power factor

(b) 0.7 lagging power factor

Solution:

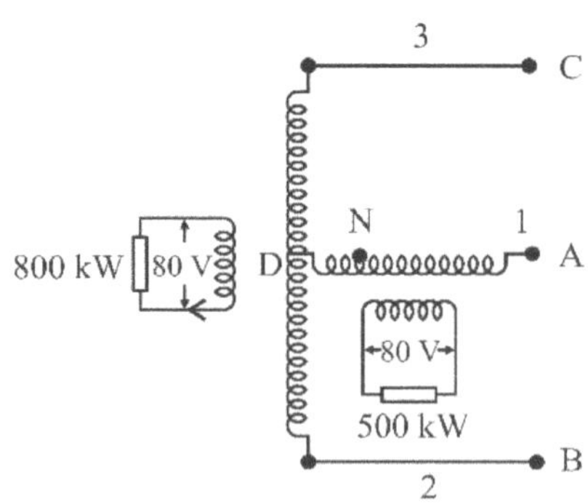

Figure 3.37 Two furnaces fed by 3-pH supply

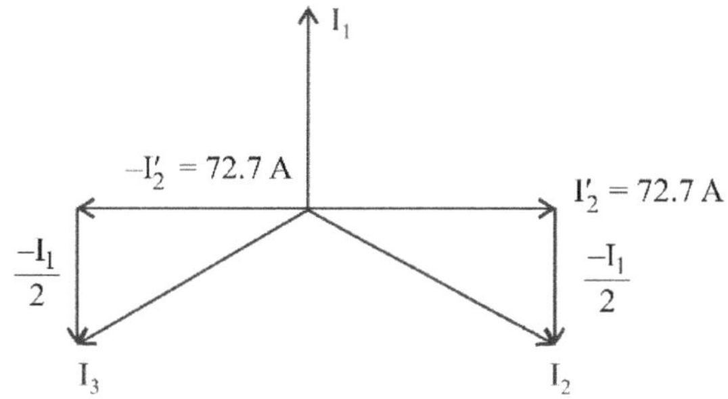

Figure 3.38 Phasor diagram

The circuit diagram for conversion from 3-phase supply of 11 kV to 2-phase supply for feeding two electric furnaces at 500 kW and 800 kW is shown in Fig.3.37.

(a) Furnaces working at unity p.f (Fig.3.38)

Transformation ratio k_1 of main transformer

$$k_1 = \frac{11000}{80} = 137.5$$

Transformation ratio k_2 of teaser transformer

$$k_2 = \frac{11000 \times 0.866}{80} = 119$$

Primary current of teaser transformer

$$I_1 = \frac{500,000}{80 \times 119} = 52.5 \text{ A}$$

Primary current of main transformer

$$I_2' = \frac{800,000}{80 \times 137.5} = 72.7 \text{ A}$$

At the tapping point D, current I_1 of teaser transformer divides equally into two halves in main winding and flows in opposite direction.

Line currents in line 2 and 3 are

$$I_2 = I_3 = \sqrt{(I_2')^2 + (I_1/2)^2} = \sqrt{72.7^2 + \left(\frac{52.5}{2}\right)^2} = 77.2 \text{ A}$$

(b) Furnaces working at 0.7 lagging p.f (Fig.3.39)

Primary current of teaser transformers

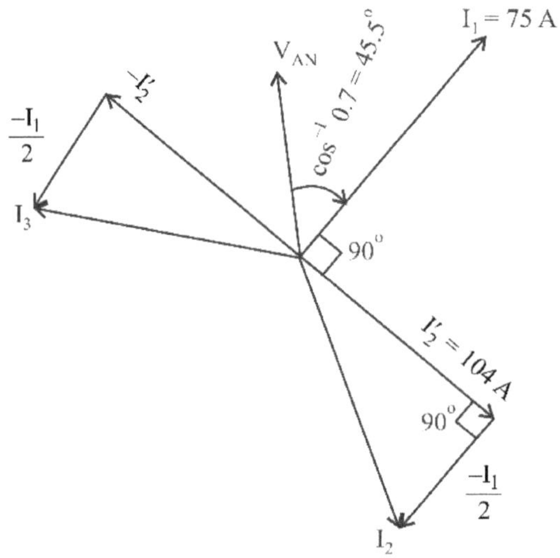

Figure 3.39 Phasor diagram

$$I_1 = \frac{500,000}{80 \times 119 \times 0.7} = 75 \text{ A}$$

Primary current of main transformer

$$I_2' = \frac{800,000}{80 \times 137.5 \times 0.7} = 104 \text{ A}$$

Current I_1 is lagging the voltage V_{AN} by $\cos^{-1}0.7$ i.e., 45.5°

Line currents in line 2 and 3 will be

$$I_2 = I_3 = \sqrt{I_2'^2 + \left(\tfrac{I_1}{2}\right)^2}$$

$$= \sqrt{104^2 + \left(\tfrac{75}{2}\right)^2}$$

$$= 110.5 \text{ A}$$

3.11 Three Phase to Six Phase Connections

1. Star or Delta / Double star
2. Star or Delta / Double delta
3. Delta / Diametral

3.11.1 Star or Delta/Double Star Connection

The primary winding is connected in star or delta and the secondary winding in double star. In Fig.3.40, transformer core having primary winding delta-connected and secondary winding in double star is shown.

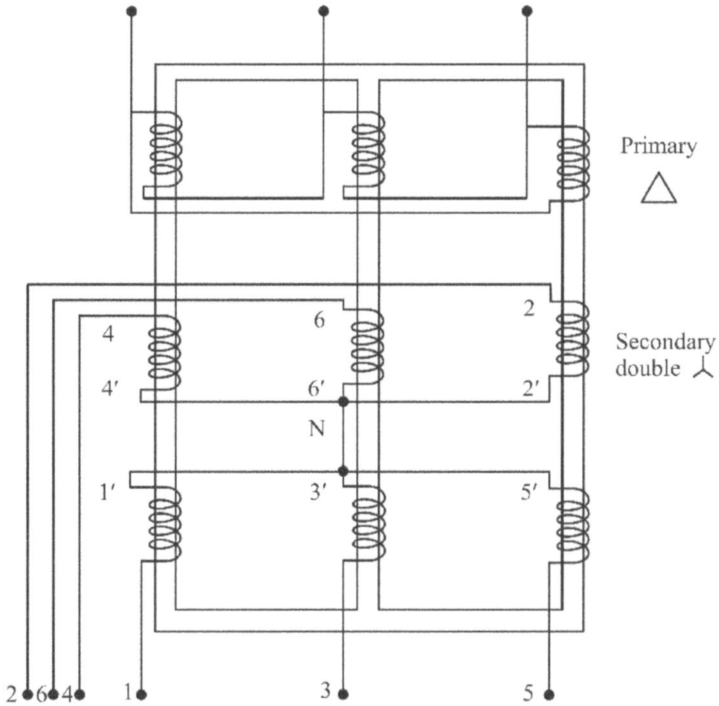

Figure 3.40 Primary winding delta connected and secondary winding double star

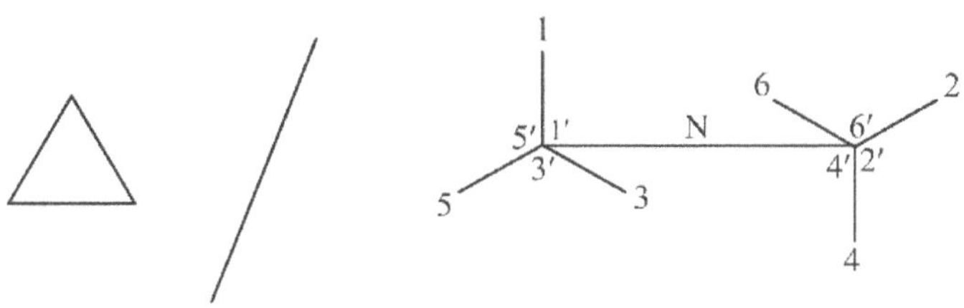

Figure 3.41 Phasor diagram of Delta/Double star connection

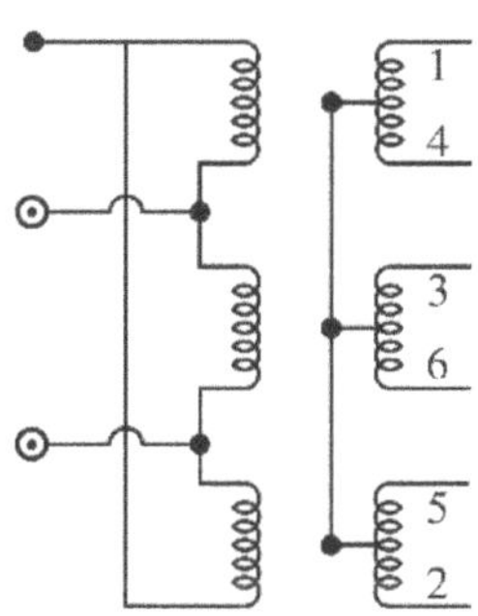

Figure 3.42 Winding representation of Figure 3.40

On a three-phase transformer core both primary and secondary windings are represented in Fig.3.40. The primary winding is delta connected and secondary winding is double star. The phasor representation is shown in Fig.3.41 and electrical representation in Fig.3.42. In Fig.3.43

secondary windings as double star 1-3-5 and 4-6-2 are shown and the vector diagram of secondary emfs is shown in Fig.3.44.

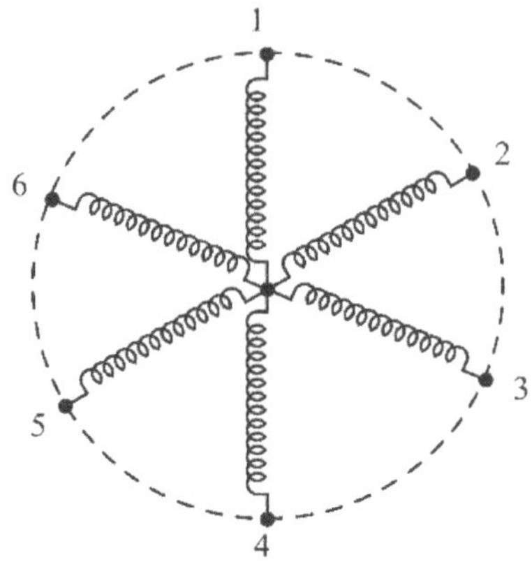

Figure 3.43 Secondary windings1-3-5 and 4-6-2 as double star

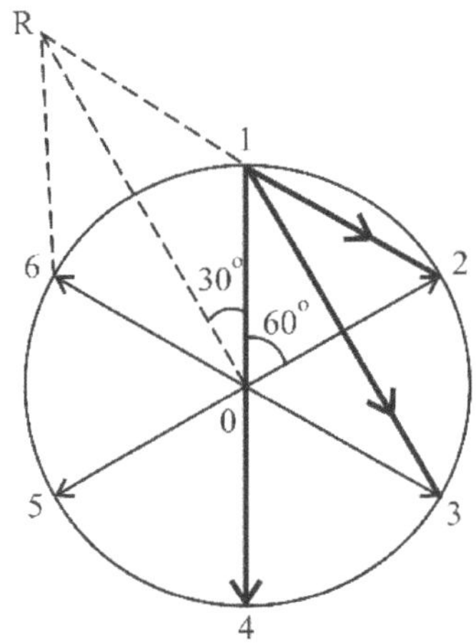

Figure 3.44 Phasor diagram of secondary emfs

Voltage between terminals 1 and 2 is

$$V_{12} = V_{10} + V_{02} = V_{10} - V_{20} = V_{10} + V_{50}$$

$$= V_{60} = V_{10} = \text{Voltage of 1phase of double star secondary}$$

Voltage between terminals 1 and 3

$$V_{13} = V_{10} + V_{03} = V_{10} - V_{30} = V_{10} + V_{60} = V_{OR}$$

$$= 2V_{10} \cos 30° = \sqrt{3}V_{10}$$

Voltage between terminals 1 and 4

$$V_{14} = V_{10} + V_{04} = V_{10} - V_{40} = V_{10} + V_{10} = 2\,V_{10}$$

3.11.2 Star or Delta / Double Delta Connection

The primary winding is connected in star or delta and the secondary winding in double delta. Winding representation of primary winding star-connected and the secondary winding in double delta is shown in Fig.3.45.

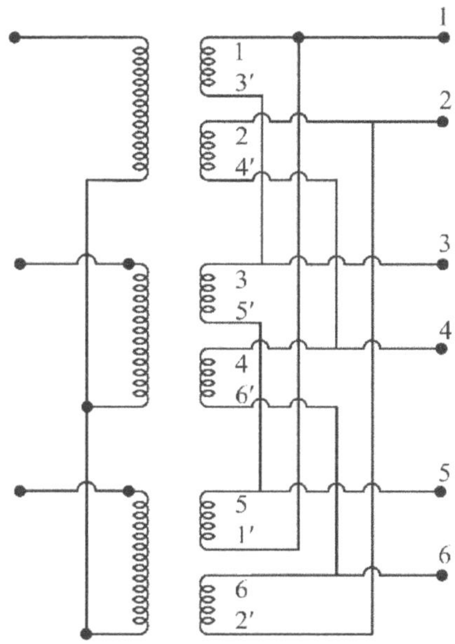

Figure 3.45 Star/Double delta connected

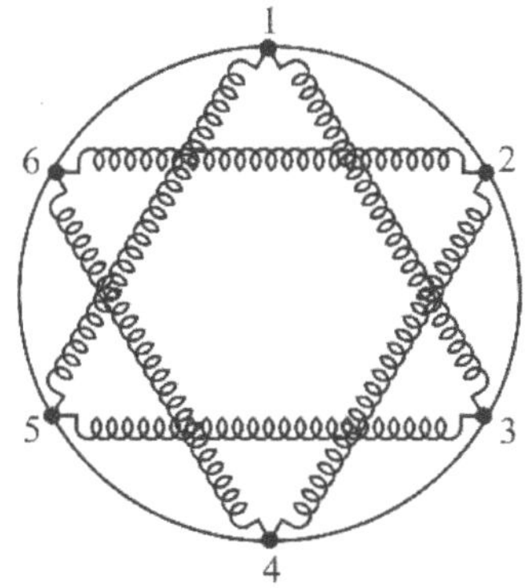

Figure 3.46 Secondary double delta connected

In Fig.3.46 secondary double delta connected winding is shown. One delta is 1-3-5 and the other is 4-6-2. The corresponding phasor diagram shown in Fig.3.47.

V_{12} = Voltage between two adjacent terminals

V_{13} = Voltage across one phase of secondary = $2V_{12}\cos 30^\circ = \sqrt{3}\,V_{12}$

V_{14} = Double the phase voltage = $2\,\dfrac{V_{13}}{\sqrt{3}} = \dfrac{2\sqrt{3}}{\sqrt{3}}V_{12} = 2V_{12}$

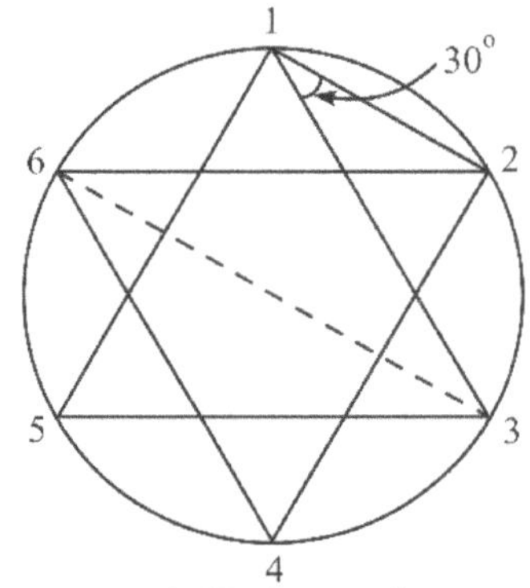

Figure 3.47 Phasor diagram

3.11.3 Delta/Diametral Connection

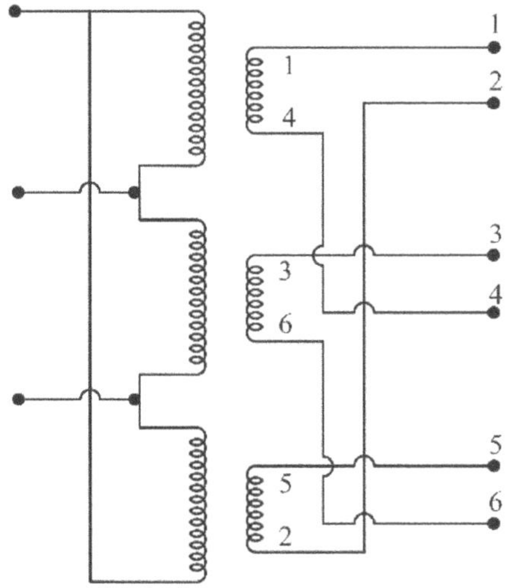

Figure 3.48 Delta/Diametral connection

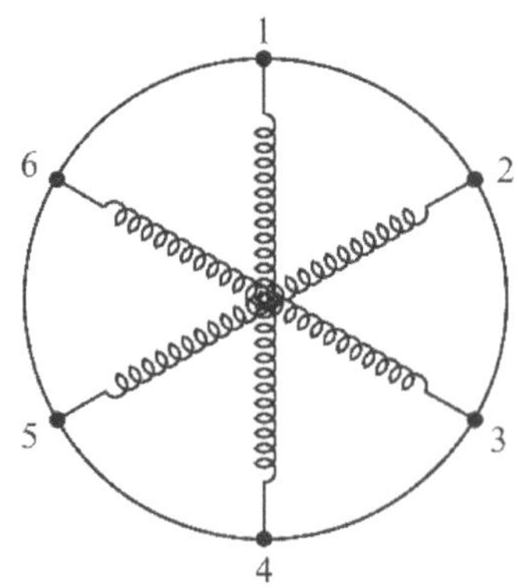

Figure 3.49 Secondary diametral connection

In Fig.3.48, the primary winding is connected in delta and secondary winding is diametral connected.

In Fig.3.49, secondary winding diametral connection is shown and the corresponding phasor diagram is shown in Fig.3.50.

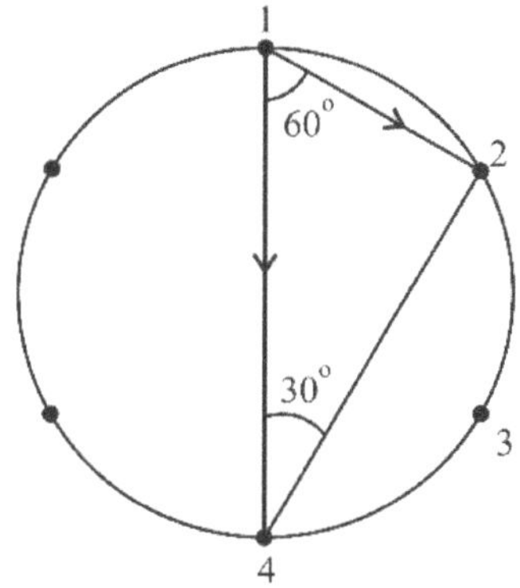

Figure 3.50 Phasor diagram

V_{12} = Voltage between adjacent terminal 1 and 2

$$= V_{14} \sin 30° = \frac{1}{2} V_{14}$$

V_{14} = Voltage of one diametral phase

V_{13} = Voltage between terminals 1 and 3

$$V_{24} = V_{14} \cos 30° = \frac{\sqrt{3}}{2} V_{14}$$

3.11.4 Zigzag Connected Transformer

Zigzag connection is used in special purpose transformers handling fairly low voltages on zigzag-connected side. A 3-phase Delta / Zigzag connected transformer is shown in Fig.3.51.

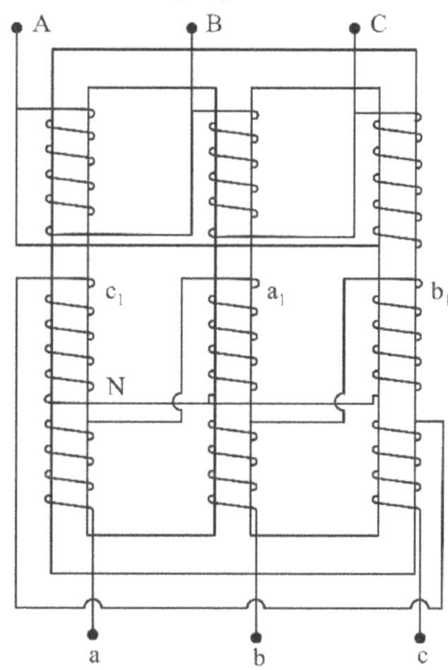

Figure 3.51 Delta/Zigzag connected transformer

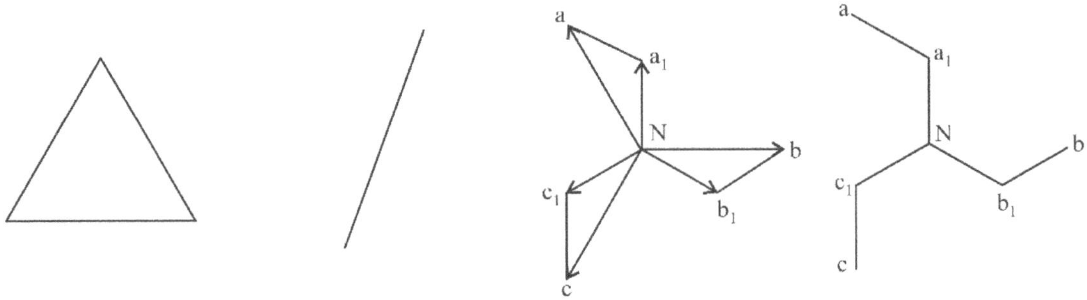

Figure 3.52 Phasor Diagram of zigzag connected winding

Each phase of zigzag connected winding on low voltage side is divided into two equal parts and those parts are placed on different limbs as shown in Fig.3.51. These two parts are so connected that their emf's are geometrically subtracted. The resultant emfs is $\sqrt{3}$ times the emf in each half section of zigzag winding as shown in Fig.3.52. The fundamental principle is that when delta connected primary winding is excited by 3-phase balanced supply, the fluxes developed in the three legs shall be 120° displaced with respect to each other. The two half sections of the zigzag winding are on adjacent legs, hence emf induced in them would also be 120° displaced.

3.12 Phasor Groups with Zigzag Connected Transformers

The demerit of zigzag connection is that the emf is reduced by about 15% as compared to normal winding. Hence the use of zigzag connected transformer is limited to low voltages where it is desired to eliminate the third harmonic emfs from the line. The following vector groups are possible with zigzag connection.

(a) **Delta–Zigzag (Dz0):** Here the *hv* winding is delta connected and *lv* winding has zigzag connection with zero-degree phase displacement.

(b) **Delta–Zigzag (Dz6):** Here the hv winding is delta connected and *lv* winding has zigzag connection with 180° phase displacement.

(c) **Star–Zigzag (Yz1):** The *hv* winding is star-connected and *lv* winding has zigzag connection with –30° phase displacement.

(d) **Star–Zigzag (Yz11):** The *hv* winding is star-connected and *lv* winding has zigzag connection with +30° phase displacement.

3.13 3 Phase to 6 Phase Conversion using Double Zigzag Connection

The primary winding is star or delta connected and the secondary has double zigzag connection as shown in Fig.3.53.

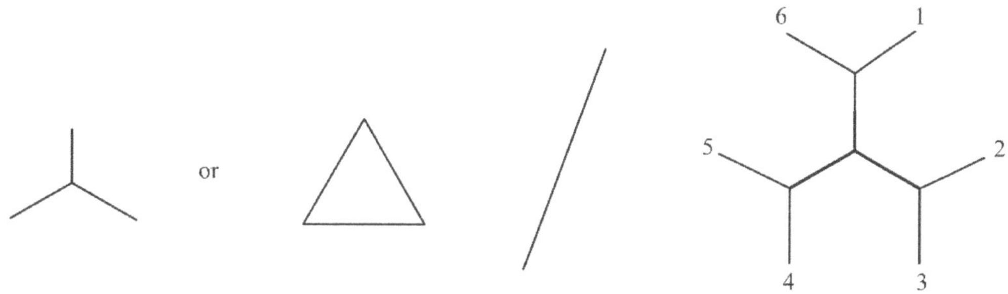

Figure 3.53 Phasor diagram of double zigzag connection

3.14 3 Phase to 12 Phase Conversion using Two Times Double Zigzag Connection

The Primary winding is star or delta connected and secondary has two times double zigzag connection as shown in Fig.3.54.

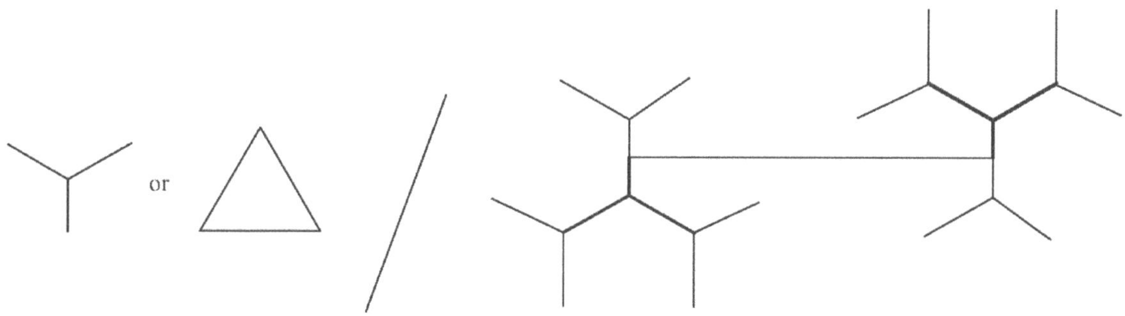

Figure 3.54 Phasor diagram of two times double zigzag connection

3.15 3 Phase to 12 Phase Conversion using Double Diametral Connection

The primary side is delta connected and on secondary side double-diametral connection is used.

3.16 Application of 3 Phase to 12 Phase Conversion

- For stepper motors higher number of phases may be needed to be supplied.
- In certain electronic circuitary higher phases supply may be required.
- For converting alternating current to direct current by synchronous converters or mercury arc rectifiers the efficiency of conversion increases by using higher number of phases.

3.17 Excitation Phenomenon and Harmonics in Transformers

The use of high flux densities in the core of transformers results in reduction of weight and hence in cost of manufacture. But this introduces serious troubles due to saturation of the magnetic circuit of the transformer which subsequently results in production of harmonics. Even when the supply voltage to a transformer is purely sinusoidal, the no load current is non-sinusoidal. The basic cause for this excitation phenomenon lies in hysteresis and saturation non-linearities of core material.

We know that the magnetizing force

$$H = \frac{NI_o}{l}\left(A/m\right) \quad .....(1)$$

where

N = Number of turns

I_o = Exciting current (A)

l = Length of flux path (m)

Thus, H is a measure of exciting current of the transformer.

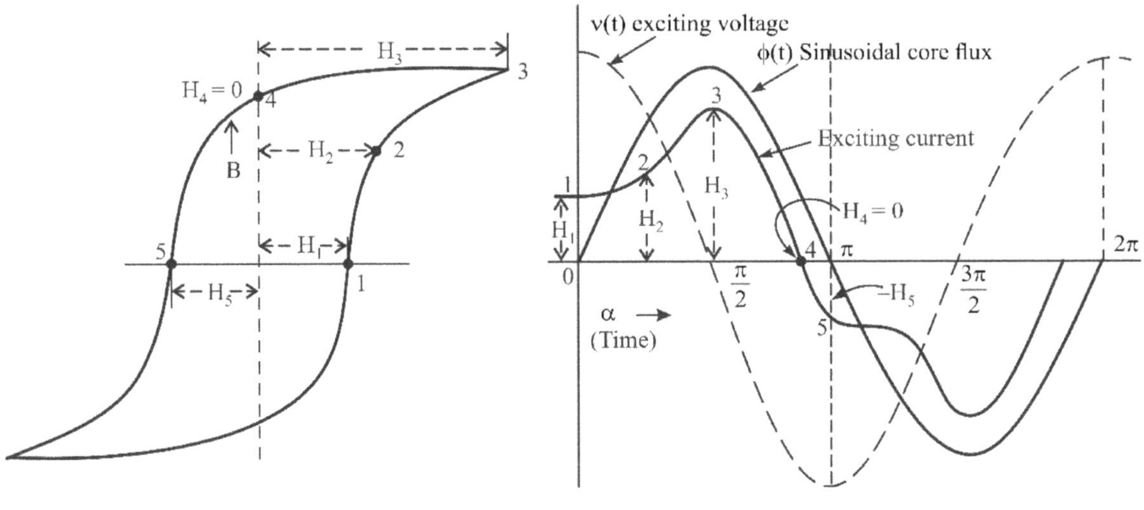

(a) Hysteresis loop (b) Exciting current waveform

Figure 3.55 Excitation phenomenon in a transformer

A simple method of plotting the variation of exciting current with time is illustrated. Let points 1, 2, 3, 4 and 5 on hysteresis loop represent flux densities at different instants in half a cycle of applied voltage. If the applied voltage is sinusoidal, the flux density and hence the flux ϕ(t) in the core must also be sinusoidal as shown in Fig.3.55(b), However, due to the shape of the hysteresis loop of transformer core material, the exciting current waveform is not sinusoidal. In Fig.3.55(a) at point 1, the flux density B is zero and the magnetizing force is H_1 which is plotted at point 1 in Fig.3.55(b). At instant of point 3, the magnetizing force is H_3 and so on. Thus, at instants of points 2, 3, 4 and 5, H_2, H_3, H_4 (= 0) and H_5 (= $-H_1$) are plotted in Fig.3.55(b) and curve is completed for the remaining half cycle. It may be noted that though the flux density in the core is sinusoidal, yet the exciting current waveform is non-sinusoidal and is highly distorted. The amount of distortion depends upon the peak value of the flux density. Further, the exciting current is advanced in phase to supply the hysteresis and eddy current losses. The exciting current wave comprises of a fundamental and series of odd harmonics represented by:

$$I_o(\alpha) = \sum_{\nu=1,3,5}^{\infty} a_\nu \sin \alpha\nu \qquad .....(2)$$

$$= a_1 \sin \alpha + a_3 \sin 3\alpha + a_5 \sin 5\alpha + ... \qquad .....(3)$$

where $$a_\nu = \frac{2}{\pi}\int_0^\pi I_o(\alpha)\sin \nu\alpha \, d\alpha \qquad .....(4)$$

As the peak value of flux density increases the amplitude of 3rd, 5th, 7th, ... harmonics also increase. The wave shape of the exciting current is non-sinusoidal and peaky. Peakiness in exciting current is due to saturation phenomenon and would be present even if hysteresis were absent.

The maxima of exciting current and flux occur simultaneously but the current zeros of exciting current advanced in time with respect to flux wave shape. As a result, the current has fundamental and odd harmonics. The strongest being the third harmonic which may be of the order of 40% of fundamental.

The no load exciting current I_o consists of magnetizing current I_μ and core loss current I_c. I_c is in phase with voltage and 90° ahead of I_μ and flux. I_c is a much smaller component and is responsible for core loss. It does not cause distortion. The magnetizing current I_μ causes distortion and introduces harmonics. I_μ is small compared to load current, hence under load conditions harmonics are hardly noticeable.

3.18 Excitation Phenomenon with Zero Hysteresis Loss

This phenomenon may be studied under two conditions.

1. Unsaturated magnetic circuit
2. Saturated magnetic circuit.

3.18.1 Unsaturated Magnetic Circuit

The waveforms are plotted as per procedure followed in Fig.3.55. In this case no harmonics are introduced in the exciting current (Fig.3.56).

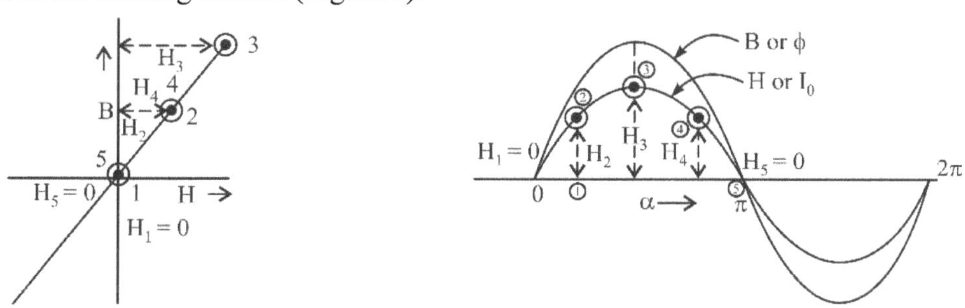

(a) Unsaturated magnetic circuit

(b) Flex and exciting current wave form are sinusoidal

Figure 3.56 Excitation phenomenon with zero hysteresis loss and linear magnetic circuit

3.18.2 Saturated Magnetic Circuit

The excitation phenomenon with saturated magnetic circuit and no hysteresis loss is shown in Fig.3.57. The wave shape of the exciting current is peaky and induces odd order of harmonics.

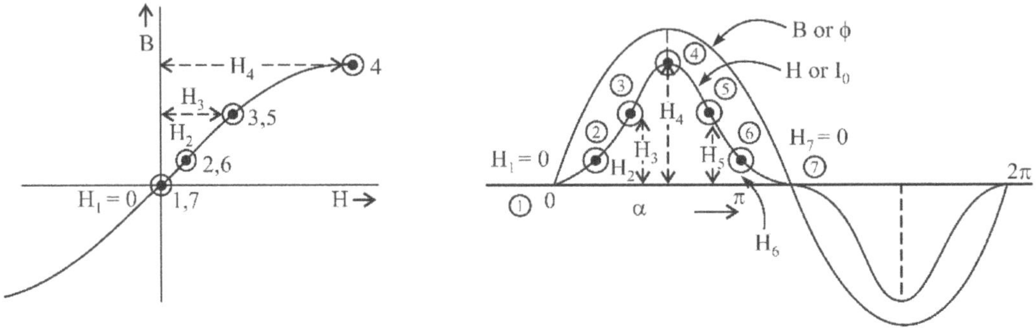

(a) Saturated magnetic circuit with no hysteresis loss

(b) Flux and exciting current wave form

Figure 3.57 Excitation phenomenon with saturated magnetic circuit

3.19 Harmonics in Delta Connected Transformer

When three 1-phase transformers are connected in delta, a voltage due to triplen harmonics gets induced before closing the mesh. When the mesh is closed, the closed mesh serves as a trap for triplen harmonic currents. All the triplen harmonics generated by the transformer exciting current form closed circulating currents around the delta. The triplen harmonics get blocked and do not circulate further. Thus, the neutral of Δ/Y and Y/Δ gets stabilized and is not allowed to float. The additional circulating currents created by the harmonics add to the ohmic power loss in the transformer.

3.20 Grounding of Transformer Neutral

In case the neutral of Y - Y transformer is not grounded, the triphen harmonics tend to force the "neutral" off its geometric centre and make it float. This results in unbalance of both phase and secondary line voltages. The output voltage waveform is also distorted. If neutral is grounded, geometric centre is maintained. Unbalanced phase currents and harmonic currents find closed path to flow. No grounding is required for $\Delta / Y, Y/\Delta$ or Δ/Δ or Δ/Δ connected transformers.

3.21 Grounding Transformer

Sometimes the neutral of a power transformer is not available for grounding (e.g., Δ–Δ). In such situations a special Y–Δ transformer is used solely for neutral grounding. This transformer is called grounding transformer. It is a step down transformer. The star-connected primaries are connected to the system and its neutral is grounded. The secondaries in Δ provide a closed path for triphen harmonic currents to circulate in them.

Under balanced conditions the current in grounding transformer is only its own exciting current. However large currents may flow in the event of single line to ground fault condition. Hence grounding transformer should be of sufficient rating.

3.22 Effect of Core Construction on Harmonic Voltages

There are two types of cases:

1. Transformers with interlinked magnetic circuits.
2. Transformers with independent magnetic circuit.

3.22.1 Transformer with Interlinked Magnetic Circuit

Three phase core type transformers fall under this category. In this case the reluctance to 3rd harmonic flux path is very high. Hence a very small value of 3rd harmonic voltage is induced. Even if Y-Y connections are used, the induced emf in the phase coils will have only a small third harmonic component and the exciting current will also free of 3rd harmonics.

3.22.2 Transformers with Independent Magnetic Circuits

To this category belong three phase shell type transformers and bank of three 1-phase transformers. The presence or absence of 3rd harmonic components in the exciting current for various connections is summarized here under.

	Connection		Presence of 3rd Harmonic in primary exciting current	
	Primary	Secondary	Line current	Phase current
1.	Y	Δ	NO	NO
2.	Δ	Y	NO	YES
3.	Δ	Δ	NO	YES
4.	Y	Y	NO	NO
5.	Y⏚ Grounded	Y⏚	YES	YES

Brief Explanation

1. With Y - Δ connection, the 3rd harmonic exciting current circulates in the secondary delta. Though there is no 3rd harmonic current on the primary side, there will also be no 3rd harmonic potential in the induced phase voltages.
2. With Δ – Y connection, the third harmonic exciting current can flow only in primary delta.
3. With Δ–Δ connection, the 3rd harmonic magnetizing current can circulate in both the primary and secondary deltas. The circulating current in primary Δ with Δ–Δ connection is less than that of Δ - Y connection.
4. With Y-Y connection, the 3rd harmonic currents will be absent both in primary lines and phases.
5. With Y⏚ - Y⏚ connection (neutral grounded) in shell type or in bank of 1-phase transformers, the induced voltages in the phases have 3rd harmonic component. However, in 3-phase core

type transformer, the 3rd harmonic voltages are suppressed and their magnitudes are comparatively small.

3.23 Effect of Harmonic Currents

- Additional I^2 R loss
- Increased core loss
- Magnetic interference with communication circuits and protective switch gear.

3.24 Effect of Harmonic Voltages

- Increased dielectric stresses
- Electrostatic interference with communication circuits.
- Resonance between inductance of transformer windings and feeder capacitance.

Remedy

- Triphen harmonics get suppressed in Δ-connected windings
- Other harmonics (5th, 7th, ...) may be suppressed by connecting resonating shunt capacitances across the transformer terminals tuned to frequencies that are to be eliminated.

3.25 Tertiary Winding (3 Winding Transformer)

The floating of neutral occurs in case of Y – Y connected transformers. Such transformers are sometimes constructed with three windings. The main windings are connected to form Y - Y connection and the third winding is used to make a closed delta connection. The third winding forming closed Δ is called Tertiary winding.

The fundamental voltages in the three phases are displaced in time by 120° from each other and may be represented as under:

$$e_{110} = E_{1m} \sin \omega t \qquad .....(1)$$

$$e_{120} = E_{1m} \sin (\omega t - 120) \qquad .....(2)$$

$$E_{130} = E_{1m} \sin (\omega t - 240) \qquad .....(3)$$

The first subscript "1" in emf represents fundamental, the second subscript 10, 20, 30 represent voltages of phases 1, 2 and 3. E_{1m} represents peak value of the fundamental emf. The instantaneous value of 3rd harmonic emfs in the three phases would be:

$$e_{310} = E_{3m} \sin 3\omega t \qquad .....(4)$$

$$e_{320} = E_{3m} \sin 3(\omega t - 120) = E_{3m} \sin 3\ \omega t \qquad .....(5)$$

$$e_{330} = E_{3m} \sin 3\ (\omega t - 240) = E_{3m} \sin 3\ \omega t \qquad .....(6)$$

where E_3m is the peak value of third harmonic emf.

Equations 4, 5 and 6 show that all the three third harmonic emfs are in phase with each other and may be represented as shown in Fig.3.58.

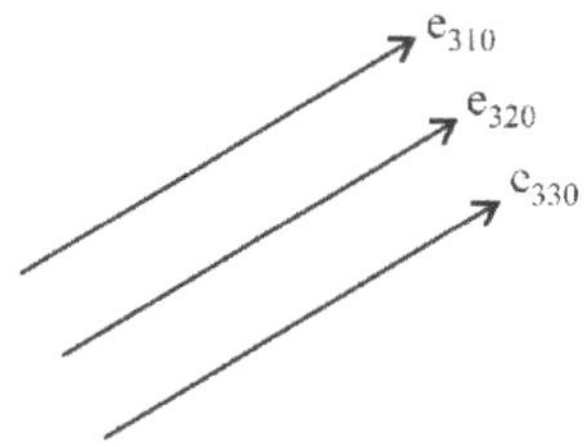

Figure 3.58 Third harmonics

Third harmonic and its odd order of multiple harmonics shall, likewise be in same phase and add up together in closed delta to circulate the triphen harmonic currents or zero sequence currents.

The winding arrangement of three winding transformer Y-Δ-Y is shown in Fig.3.59. For clarity the three winding are shown separately. The terminals of star-connected primary winding are A, B & C and that of secondary winding a, b, c. The tertiary winding is shown in Fig.3.59 connected in delta. Fig.3.60 gives vector representation.

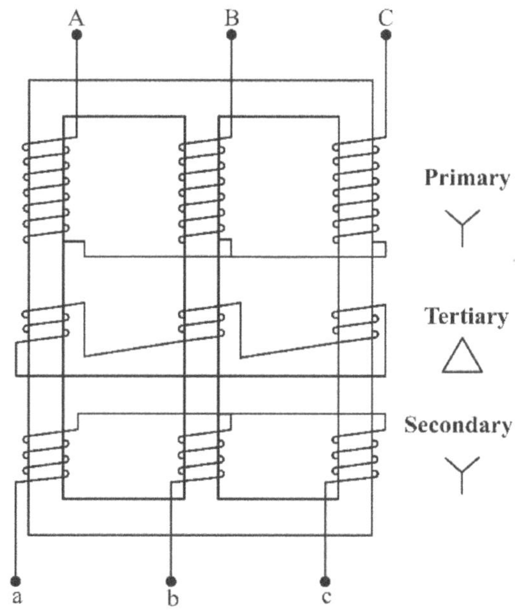

Figure 3.59 Three winding transformer

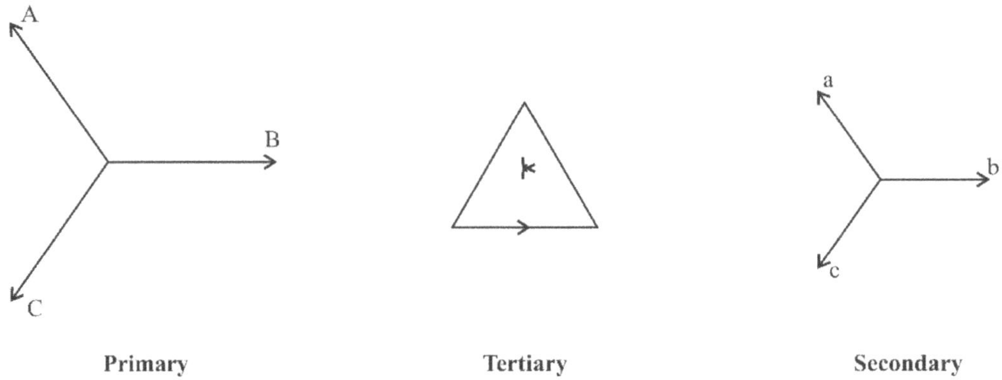

Figure 3.60 Vector representation of three winding transformer

The tertiary winding may serve any of the following purposes:

(a) To suppress harmonic voltages and allow triplen harmonic currents to circulate in closed Δ to stabilize the neutral in star star connected transformers.

(b) To allow sufficient earth fault current to flow for operation of protective devices.

(c) To supply an additional load at a voltage different from that of primary and secondary voltages.

(d) To feed two outgoing transmission system at different voltages at a generating power station.

(e) To subdivide the secondary load into two separate windings for reducing short circuit current.

(f) Delta connected tertiary winding limits voltage imbalance when the load is unbalanced.

(g) Delta connected tertiary winding reduces the impedance offered to zero sequence currents thereby allowing a larger earth fault current to flow for proper operation of protective equipment.

(h) To supply capacitor bank for power factor improvement.

3.26 Stabilization by Tertiary Winding

Let one phase of the secondary winding be loaded Fig.3.61 and a load current I_2 to flow through it. The secondary load current I_2 shall induce a current I_3 (zero sequence current) in tertiary winding. The secondary current I_2 will also induce current I_1 in phases B and C.

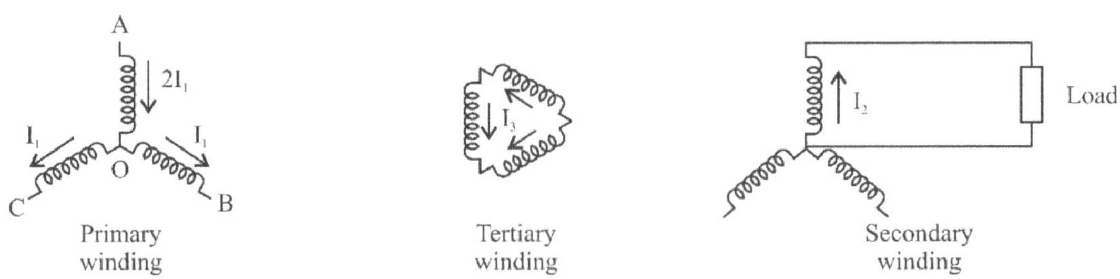

Figure 3.61 Three winding transformer neutral not ground

On applying kirchoff's law at point O, the current in phase A is $2I_1$ as the zero-sequence current in phase A and B are in same phase. To obtain balance

$$I_1 = I_3$$

$$I_2 = 2I_1 + I_3 = 3I_1 \qquad \therefore \quad I_3 = \frac{I_2}{3} = I_1$$

Hence the current I_3 in tertiary winding is one-third of the secondary current I_2. The output load or fault current flows through one phase only of the secondary whereas the corresponding input load current in absence of primary neutral, has to return through the other two phases B & C (Fig.3.61), which have very high reactances owing to the absence of their secondary currents. The voltage in phases B and C rise and become nearly equal to the primary line voltage whereas the voltage of the faulty or loaded phase A reduces to a very low value. Thus, a serious voltage unbalance is developed. By use of Δ-connected tertiary winding zero sequence currents provide low reactance path and control the voltage imbalance under unbalanced load conditions and also on occurrence of ground faults.

3.27 Equivalent Circuit of 3 Winding Transformer

Three winding transformer may be a three single phase or a single three phase transformer. If analysis is done for a single phase transformer, the same procedure can be followed for three phase transformers, keeping in mind that the power of three phase transformer would be three times the power of single phase transformers.

The equivalent circuit of a three winding transformer is given in Fig.3.62.

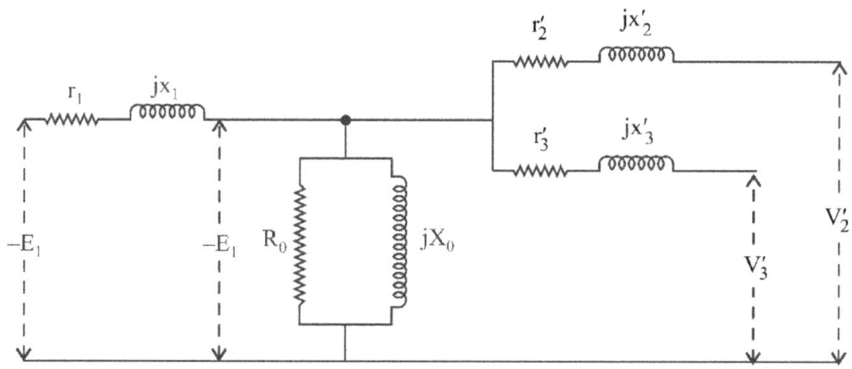

Figure 3.62 Equivalent circuit of a 3 winding transformer

The equivalent circuit has been made considering one primary winding and two secondary winding. Let r_1, r_2 and r_3 be resistances or primary, secondary and tertiary (second secondary winding) and corresponding leakage reactances be x_1, x_2 and x_3. In the equivalent circuit of Fig.3.62 the resistances and leakage reactances of secondary and tertiary windings are referred to primary side. The total equivalent resistances and leakage reactances for three windings are given as under (considering approximate equivalent circuit).

$$R_1 = r_1 + r_2' = r_1 + \left(\frac{N_1}{N_2}\right)^2 r_2$$

$$R_2 = r_1 + r_3' = r_1 + \left(\frac{N_1}{N_3}\right)^2 r_3$$

$$R_3 = r_2' + r_3' = \left(\frac{N_1}{N_2}\right)^2 r_2 + \left(\frac{N_1}{N_3}\right)^2 r_3$$

$$X_1 = x_1 + x_2' = x_1 + \left(\frac{N_1}{N_2}\right)^2 x_2$$

$$X_2 = x_1 + x_3' = x_1 + \left(\frac{N_1}{N_3}\right)^2 x_3$$

$$X_3 = x_2' + x_3' = \left(\frac{N_1}{N_2}\right)^2 x_2 + \left(\frac{N_1}{N_3}\right)^2 x_3$$

where N_1, N_2 and N_3 are number of turns per phase on primary, secondary and second secondary windings. The values of core loss component of resistance R_o and magnetizing reactance X_o may be obtained by performing the open circuit test keeping both the secondary windings open.

The values of R_1, X_1, R_2, X_2 and R_3 and X_3 are obtained by performing short circuit test three times keeping one winding open every time.

3.28 Parallel Operation of Transformers

To connect two or more transformers in parallel for sharing the total load is termed parallel operation of transformers. There are certain conditions to be satisfied for connecting single phase and three phase transformers in parallel. The conditions are classified as:

1. Essential conditions
2. Desirable conditions

Essential conditions are those without fulfilling which the transformers cannot be connected in parallel whereas desirable conditions are those that even if we do not fulfill them, the transformers may be connected in parallel. But in that case, there may be circulating current and the transformers may not be loaded to their rated capacities.

3.29 Parallel Operation of Single Phase Transformers

Essential Conditions

There are two essential conditions which must be fulfilled to connect two transformers in parallel.

1. Same frequency
2. Same polarity

Desirable Conditions

1. Same Voltage Ratio: The two transformers should have the same ratio of turns in their primary and secondary windings otherwise there will be a circulating current between them.
2. Same per unit impedance.
3. Same resistance to leakage reactance ratio in case the above two conditions are not satisfied. The load kVA can never be equal to the sum of kV as supplied by the two transformers. In that case one of the two transformers cannot be loaded to its full rating.

3.30 Per Unit System

If resistance and leakage reactance of the transformers are given in pu systems, then they should be transferred to the base of one of the transformers in the following way.

Let P_A and P_B be the ratings of the two transformers and in per unit system their impedances be $Z_A = R_A + j\ X_A$ and $Z_B = R_B + j\ X_B$. Changing the impedance of second transformer to the base kVA of the first transformer, we have

PU resistance of 2nd transformer to the base of 1st transformer $= \frac{P_A}{P_B} . R_B$

PU leakage reactance of 2nd transformer to the base of 1st transformer $= \frac{P_A}{P_B} . X_B$

3.31 Parallel Operation of 3 Phase Transformers

3.31.1 Essential Conditions

1. Same frequency: Both the transformers must operate at the same frequency.
2. Same phase sequence: The two transformers should have same phase sequence. The anti-clockwise direction of rotation of vectors is taken as positive. In Fig.3.63 correct phase sequence of two transformers is indicated. Both have anti clockwise direction of rotation. In Fig.3.64 incorrect phase sequence of second transformer is indicated. The phase sequence may be identified by a phase sequence indicator.

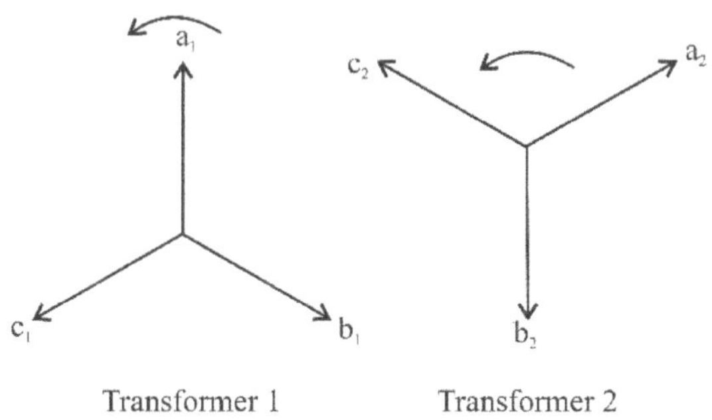

Figure 3.63 Correct phase sequence

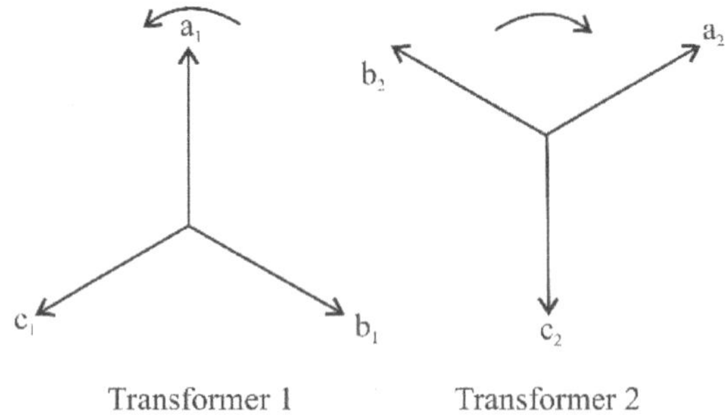

Figure 3.64 Incorrect phase sequence

If the phase sequence is incorrect, interchange any two terminals of the second transformer.

3. Zero relative phase displacement: There should be zero relative phase displacement between the corresponding phases of two transformers. In Fig.3.64 the phase sequence of two transformers is same, but there is relative phase displacement between the corresponding phases. So, voltage will exist between a_1a_2, b_1b_2 and c_1c_2 and hence current will flow if connected in parallel. Hence prior to connecting them in parallel relative phase displacement must be made zero.

3.31.2 Desirable Conditions

The desirable conditions are same as mentioned for single-phase transformers and summarized below.

1. Same voltage ratio.
2. Same per unit impedance.
3. Same resistance to leakage reactance ratio.

3.32 Load sharing by Two Transformers with Equal Voltage Ratio

Two transformers A and B with same voltage ratio are connected in parallel as shown in Fig.3.65. Their approximate equivalent circuit referred to the secondary side is shown in Fig.3.66.

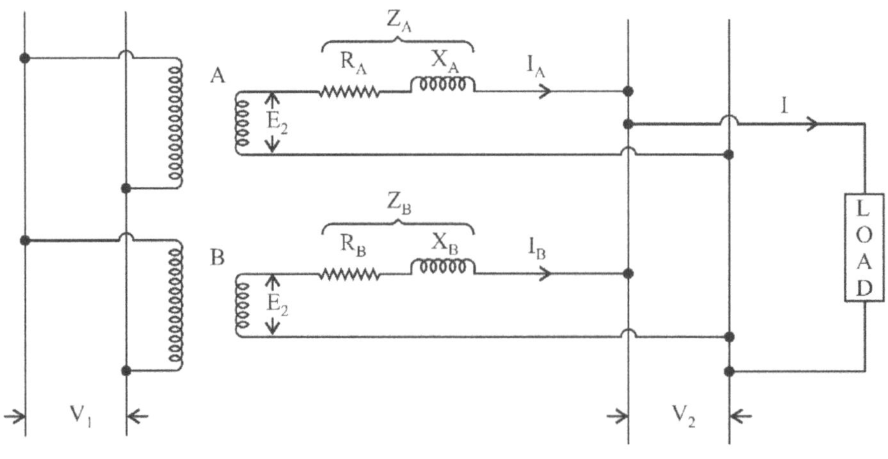

Figure 3.65 Two transformers connected in parallel

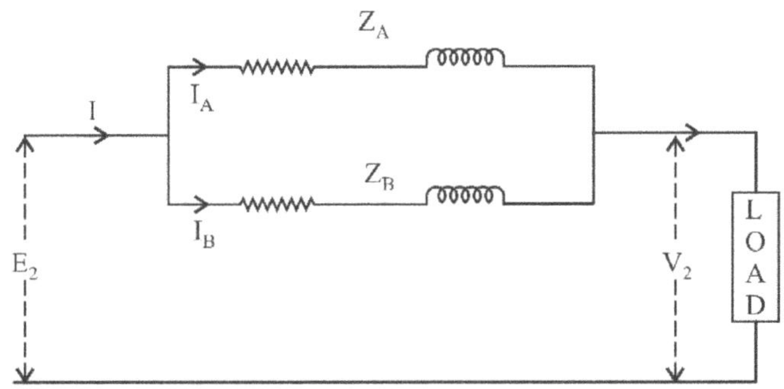

Figure 3.66 Equivalent circuit

Let $Z_A = R_A + j X_A$ (Impedance of transformer A)

$Z_B = R_B + j X_B$ (Impedance of transformer B)

R_A = Equivalent resistance of transformer A

X_A = Equivalent leakage reactance of transformer A

R_B = Equivalent resistance of transformer B

X_B = Equivalent leakage reactance of transformer B

V_1 = Primary supply voltage (V)

E_2 = Induced emf on secondary side (V)

V_2 = Load voltage on secondary side (V)

I_A = Current supplied by transformer A (A)

I_B = Current supplied by transformer B (A)

I = Total load current = $I_A + I_B$

Total impedance drop is equal to the voltage drop of each transformer in equivalent circuit of Fig.3.66.

$$I_A Z_A = I_B Z_B = I Z_{AB}$$

where Z_{AB} = Equivalent impedance of transformers A and B in parallel

$$= \frac{Z_A Z_B}{Z_A + Z_B}$$

and, $I = I_A + I_B$

From above equations,

$$I_A = \frac{Z_B}{Z_A + Z_B} I$$

$$I_B = \frac{Z_A}{Z_A + Z_B} I$$

Let P be the total load and P_A and P_B be the loads shared by transformers A and B respectively, then

$$P = V_2 I \times 10^{-3} \text{ kVA}$$

$$P_A = V_2 I_A \times 10^{-3} \text{ kVA}$$

$$= V_2 \frac{Z_B}{Z_A + Z_B} I \times 10^{-3} \text{ kVA}$$

$$= \frac{Z_B}{Z_A + Z_B} P \text{ kVA}$$

and,

$$P_B = V_2 I_B \times 10^{-3} \text{ kVA}$$

$$= V_2 \frac{Z_A}{Z_A + Z_B} I \times 10^{-3} = \text{kVA}$$

$$\frac{Z_A}{Z_A + Z_B} . P$$

From equations 7 and 8

$$\frac{P_A}{P_B} = \frac{Z_B}{Z_A}$$

Hence the load shared by two transformers is inversely proportional to their impedances. The phasor diagram is shown in Fig.3.67.

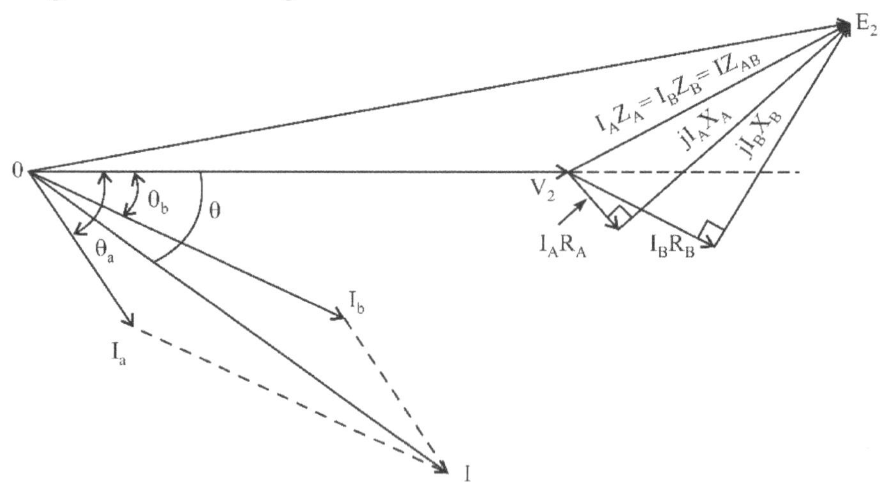

Figure 3.67 Phasor diagram

V_2 represents the common terminal voltage. I_A and I_B drawn at load power factor angles of the two transformers. The resultant of I_A and I_B gives total load current I. The no load secondary voltage E_2 is obtained by adding the impedance drop $I_A Z_A = I_B Z_B = IZ_{AB}$ to the terminal voltage V_2 the load power factor is cos θ.

Example 3.2:

Two transformers each of 600 kVA are connected in parallel. One has a resistance and leakage reactance of 1% and 4% and the other has 1.5% and 6% respectively. Determine the load shared by each transformer and the corresponding power factor when the total load shared by the two transformers is 1000 kVA at 0.8 power factor lagging.

Solution:

Since the rating of both transformers is same, the percentage impedances of the two transformers will remain same as given in the problem.

Here $\% Z_A = 1 + j4$

$\% Z_B = 1.5 + j6 = 1.5\ (i + j4)$

$\% (Z_A + Z_B) = 2.5 + j10 = 2.5(1 + j4)$

Total load $P = 1000\ (\cos \phi)$

$= 800$ kW

Load shared by transformer A

$$P_A = \frac{Z_B}{Z_A + Z_B}.P$$

$$= \frac{1.5(1 + j4)}{2.5(1 + j4)} \times 800 \text{ kW}$$

$$= 480 \text{ kW at 0.8 pf lagging}$$

and $$P_B = \frac{Z_A}{Z_A + Z_B}.P = \frac{1 + j4}{2.5(1 + j4)} \times 800 \text{ kW}$$

$$= 320 \text{ kW at 0.8 pf lagging}$$

3.33 Load sharing by Two Transformers with Unequal Voltage Ratio

Consider two transformers A and B running in parallel with unequal voltage ratios. Let E_A and E_B be the no load secondary emfs of the two transformers.

Let Z_A = Impedance of transformer A

$= R_A + jX_A$

Z_B = Impedance of transformer B

$= R_B + jX_B$

I_A = Current supplied by transformer A

I_B = Current supplied by transformer B

I = Load current

V_2 = Load voltage

Z = Load impedance

The equivalent circuit referred to secondary side is shown in Fig.3.68.

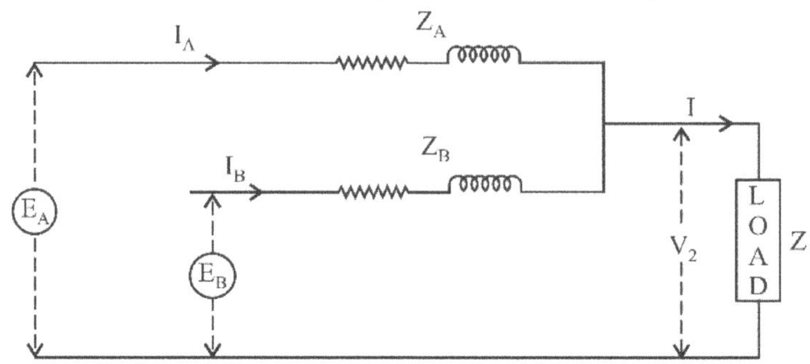

Figure 3.68 Equivalent circuit of two transformers operating in parallel with unequal secondary no load voltages

Applying Kirchoff's law to the circuit of transformer A,

$$I_A = \frac{E_A - V_2}{Z_A}$$

Similarly, for circuit of transformer B

$$I_B = \frac{E_B - V_2}{Z_B}$$

Load current $$I = \frac{V_2}{Z}$$

Further the load current

$$I = I_A + I_B$$

Putting the values of I_A, I_B and I from equations 1, 2 and 3

$$\frac{V_2}{Z} = \frac{E_A - V_2}{Z_A} + \frac{E_B - V_2}{Z_B}$$

or $$V_2\left[\frac{1}{Z} + \frac{1}{Z_A} + \frac{1}{Z_B}\right] = \frac{E_A}{Z_A} + \frac{E_B}{Z_B}$$

Representing Z, Z_A, Z_B in terms of their admittances

$$Y = \frac{1}{\dot{Z}};\ Y_A = \frac{1}{\dot{Z}_A},\ Y_B = \frac{1}{\dot{Z}_B}$$

Putting these values in equation 5, we get

$$V_2(Y + Y_A + Y_B) = E_A Y_A + E_B Y_B$$

or $$V_2 = \frac{E_A Y_A + E_B Y_B}{Y + Y_A + Y_B}$$

Putting the value of V_2 in equations 1 and 2 we obtain I_A and I_B and from equation 4 we get total load current. Power factors of both transformers and their load shared can be determined easily.

The Phasor diagram is shown in Fig.3.69.

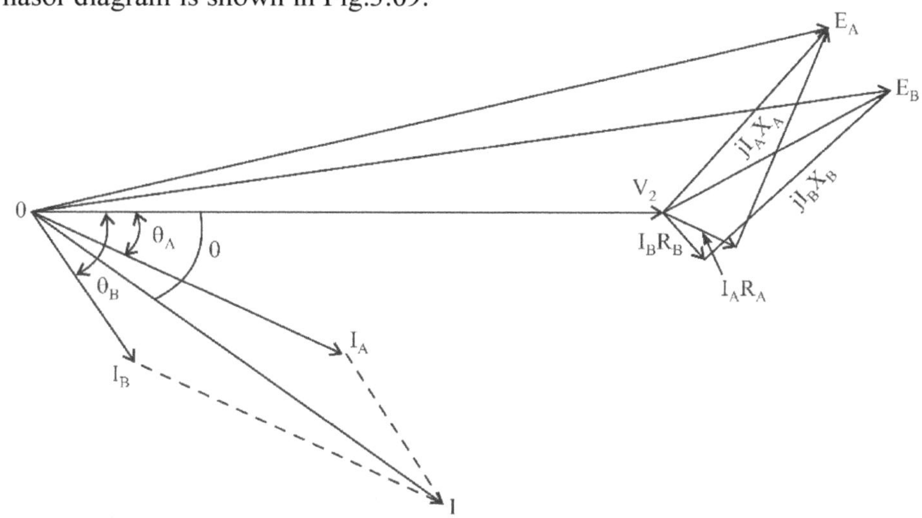

Figure 3.69 Phasor diagram

- I_A and I_B are drawn at lagging pf angles θ_A and θ_B from terminal voltage V_2.
- $I_AR_A + j\, I_A\, X_A$ drops are drawn.
- Voltage E_A is the vector sum of V_2 and I_AR_A plus $j\, I_A\, X_A$ drop.
- $I_B\, R_B + j\, I_B\, X_B$ drops are drawn.
- Voltage E_B is the vector sum of V_2 plus $j\, I_B\, X_B$ drop.
- Vector sum of I_A and I_B gives the load current I and angle θ from V_2 gives the load pf.

In this case there will be a circulating current I_C, which is given by

$$I_C = \frac{E_A - E_B}{Z_A + Z_B}$$

Example 3.3:

A 400 kVA transformer with 1% resistance and 4% leakage reactance is connected in parallel with 200 kVA transformer having 1.5% resistance and 5% leakage reactance. The secondary terminal voltage is 400 V on load. Determine how they share a load of 500 kVA at 0.8 pf lagging.

Solution: Since the ratings of two transformer are not the same. We must change the impedance of the second transformer on base kVA of first transformer.

$$\% \text{ resistance of } 2^{nd} \text{ transformer} = \frac{P_A}{P_B} \times 1.5 = \frac{400}{200} \times 1.5 = 3\%$$

$$\% \text{ leakage reactance of } 2^{nd} \text{ transformer} = \frac{P_A}{P_B} \times 5 = \frac{400}{200} \times 5 = 10\%$$

$$Z_A = (1 + j4)\%$$

$$Z_B = (3 + j10)\%$$

$$Z_A + Z_B = (4 + j14)\%$$

Given load $S = 500 (\cos \phi - j \sin \phi)$

$$= 500 (0.8 - j\,0.6)$$

$$= 400 - j\,300$$

Load supplied by I^{st} transformer

$$S_A = \frac{Z_B}{Z_A + Z_B} S_A = \frac{Z_B}{Z_A + Z_B} S = \frac{3 + j10}{4 + j14} (400 - j300)$$

$$= 284 - j219 = \sqrt{284^2 + 219^2} = 359 \text{ kVA}$$

At power factor $\cos \phi_A = \frac{284}{359} = 0.793$ Lag

Load supplied by 2nd transformer

$$S_B = \frac{Z_A}{Z_A + Z_B} S = \frac{1 + j4}{4 + j14} (400 - j\,300)$$

$$= 116 - j\,81 = \sqrt{116^2 + 81^2} = 142 \text{ kVA}$$

At p.f $\cos \phi_B = \frac{116}{142} = 0.82$ lag

3.34 Auto-Transformer

We have both single-phase and three-phase Auto-transformers. An auto-transformer has only one winding which serves the purpose of both primary and secondary windings.

3.34.1 Single-phase Auto-Transformer

As mentioned above there is only one winding placed on a closed magnetic circuit of low reluctance as shown in Fig.3.70(a) and its equivalent electric representation is given in Fig.3.70(b). A section of winding is common to both primary and secondary. In theory and operation, an auto-transformer is similar to a two winding transformer where primary and secondary sides are interchangeable.

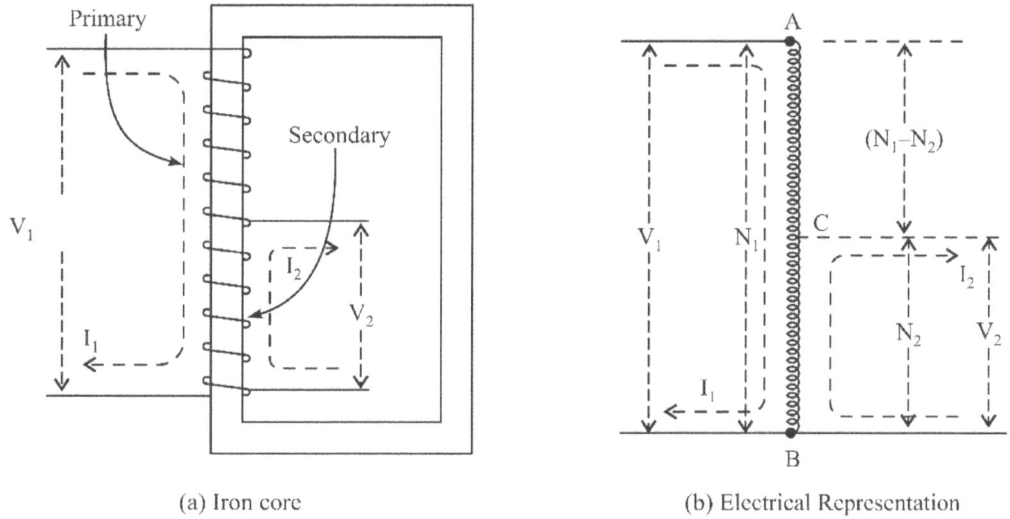

Figure 3.70 Auto-transformer

Let the number of turns in primary winding be N_1 and in secondary winding N_2. I_1 and I_2 be the primary input and secondary output current. Ignoring no load current and core losses, the transformation ratio is

$$k = \frac{V_2}{V_1} = \frac{N_2}{N_1} = \frac{I_1}{I_2}$$

3.34.2 Ratio of Weight of Copper in Auto-Transformer to Two Winding Transformer

For same voltage ratio and rating an auto-transformer requires less weight of copper than what is required by a normal two winding transformer.

Weight of copper in winding $\propto$ volume of copper

$\propto$ Length of conductor × Cross-section

$\propto$ Number of turns × current

Since cross-section of conductor is decided by the current carried by it. A simple relationship is

$$\text{Cross section} = \frac{\text{Current carried}}{\text{Permissible current density}}$$

Normal Two Winding Transformers

Weight of copper in primary winding $\propto N_1 I_1$

Weight of copper in secondary winding $\propto N_2 I_2$

Total weight of copper $\propto (N_1 I_1 + N_2 I_2)$

$\propto 2 N_1 I_1$

As for an ideal transformer $N_1I_1 = N_2 I_2$

Auto-Transformer

From Fig.3.70(b), current in section AC having turns $(N_1 - N_2)$ is I_1

Current in section BC having N_2 turns is $(I_2 - I_1)$

Total weight of copper in both sections' AC and BC

$$\propto [(N_1 - N_2)I_1 + N_2(I_2 - I_1)]$$

$$\propto [N_1I_1 - N_2I_1 + N_2I_2 - N_2I_1]$$

$$\propto 2[N_1I_1 - N_2I_1]$$

$$\text{Ratio of weight of copper} = \frac{\text{Weight of copper in auto transformer}}{\text{Weight of copper in normal transformer}}$$

$$= \frac{2(N_1I_1 - N_2I_1)}{2N_1I_1}$$

$$1 - \frac{N_2}{N_1} = 1 - k$$

If the voltage transformation ratio k is very small, then $(1 - k)$ is approximately equal to 1, so the weight of copper in auto-transformer will be nearly equal to the weight of two winding transformer. In such case it is preferable to use a normal two winding transformer.

If k is large say 0.4 and above, then there will be sufficient saving of copper in an auto-transformer.

3.43.3 Power Transferred by Transformer Action in Auto-Transformer

In a normal two winding transformer the entire power is transferred by transformer action from primary to secondary side. But in an auto-transformer only a part of power is transferred from transformer action and the rest is conducted directly.

Power transferred in auto-transformer by transformer action Fig.3.70(b)

$$= \text{Voltage across section BC} \times \text{Current in it}$$

$$= V_2 (I_2 - I_1)$$

$$= V_2I_2 \left(1 - \frac{I_1}{I_2}\right)$$

$$= V_2 I_2 (1 - k)$$

For an ideal transformer $V_2 I_2 = V_1 I_1$

Hence power transferred by transformer action

$$= V_1I_1 (1 - k)$$

$$= \text{Input power } (1 - k)$$

Power conducted directly

$$= \text{Power input} - \text{Power transferred by transformer action}$$

$$= V_1 I_1 - V_1 I_1 (1 - k)$$

$$= V_1 I_1 \times k$$

$$= k \times \text{Power input}$$

3.35 Three Phase Auto-Transformer

A 3-phase auto transformer is shown in Fig.3.71.

Let N_1 = No. of turns in primary/ph

N_2 = No. of turns in secondary/ph

Current in winding A_a, B_b, $C_c = I_1$

Current in winding oa, ob & oc $= I_2 - I_1$

$$V_1 \text{ phase} = \frac{V_1}{\sqrt{3}}$$

$$V_2 \text{ phase} = \frac{V_2}{\sqrt{3}}$$

Transformation ratio $$k = \frac{V_2 \text{phase}}{V_1 \text{phase}} = \frac{N_2}{N_1} = \frac{I_1}{I_2}$$

Ratio of weight of copper and power transferred by transformer action can be found out in a similar manner as done for single-phase auto transformer.

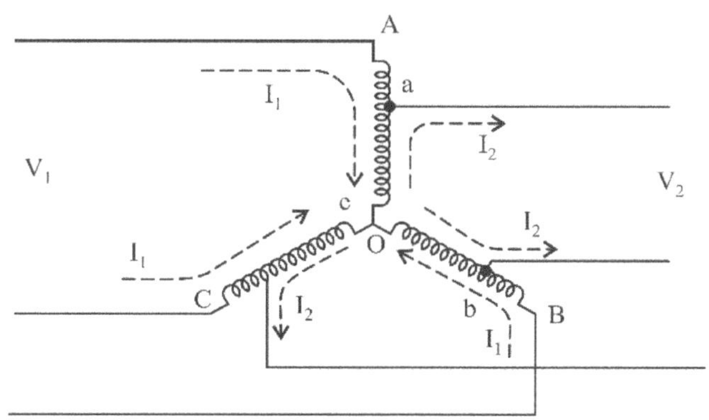

Figure 3.71 Three-Phase Auto-transformer

3.36 Two Winding Transformer as Auto-Transformer

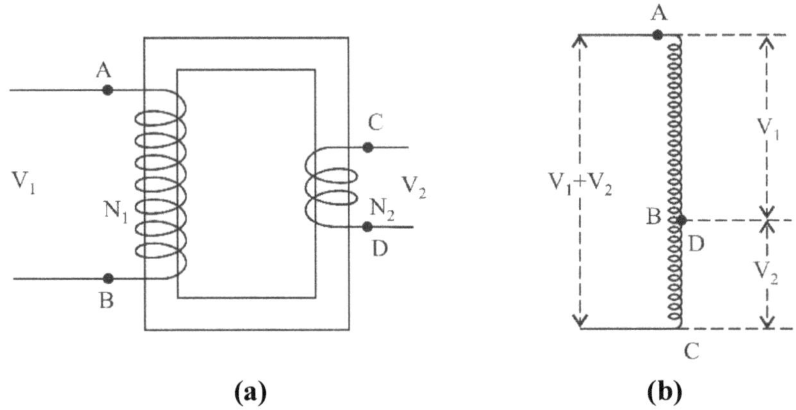

Figure 3.72 Two winding transformer as auto transformer

For making a two-winding transformer of Fig.3.72(a) as an auto-transformer of Fig.3.72(b), join the terminals B and D and apply a voltage $V_1 + V_2$ across AC, the transformed voltage across CD shall be V_2.

$$\frac{V_2}{V_1 + V_2} = \frac{N_2}{N_1 + N_2}$$

Also, the transformed voltage across AD should be

$$\frac{V_1}{V_1 + V_2} = \frac{N_1}{N_1 + N_2}$$

Other ratios of voltage transformation can be obtained by changing polarities.

3.37 Efficiency

The efficiency of an auto-transformer may also be determined in a similar manner as done for two winding transformers.

3.38 Merits

- Some saving of copper
- Somewhat higher efficiency
- Continuously varying voltage may be obtained
- Smaller size and lower cost

Demerits

- Direct connection of *hv* and *lv* sides
- Short circuit current is larger as compared to a two-winding transformer

3.39 Applications

- Used as a regulating transformer
- Used as a booster to raise the voltage in alternating current feeder.

REVIEW QUESTIONS

1. Give difference between a three phase three limb and a three phase five limb transformer.
2. What happens in a three phase shell type transformer when the applied voltage is reversed in the middle winding?
3. Compare a three phase unit transformer with a bank of three single phase transformers.
4. Discuss various connections of Three Phase transformers.
5. What is understood by phasor Groups? How are zero phase displacements obtained?
6. What is the need of parallel operation of three phase transformers?
7. What is open Delta Connection and how much is the de-rated power is available?
8. Explain Scott-connection.
9. Discuss about Star/Double Delta connection. What would be the relation of voltage between two adjacent terminals and one phase of secondary of this transformer?
10. Explain zigzag connected transforms and its application.
11. Discuss excitation phenomenon in transformers. How does the exciting current waveform become distorted even when supply voltage is purely sinusoidal?
12. What is grounding transformer and its function?
13. How does core construction affects the harmonic voltages in a transformer?
14. What is Tertiary winding and its purpose?
15. Give equivalent circuit of a three winding transformer.
16. What is meant by Parallel operation of Three Phase Transformers and what are the conditions to be satisfied?
17. Give working and construction of a three phase Auto-Transformer. Find the ratio of copper in Auto-transformer to Normal Transformer.
18. How to use a Two Winding Transformer as an Auto-Transformer? What are the different voltage transformation ratios obtainable?

PROBLEMS

1. A 3-phase, Delta/Star connected transformer has voltage ratio 20000/2000 V at 50 Hz. Each core has a net iron section of 504 cm^2 and flux density of 1.2 T. Determine the number of turns per phase on primary and secondary windings.

   [Ans: 1490, 86]

2. A 3-phase transformer has 500 turns on primary and 50 turns on secondary. The supply voltage is 3300 V. Determine secondary voltage on no load when the windings are connected in (a) Delta/Star (b) Stat/Delta.

   [Ans. 571.6V, 190.5V]

3. A 22000/400 V, 50 Hz Delta/Star connected, 3-phase transformer supplies a balanced load at 0.8 pf lagging. The line current on primary side in 5 A. Determine primary phase current, secondary line current and the output power.

   [Ans. 2.89 A, 275 A, 15.24 kV]

4. A 120 kVA, 6000/400 V, star/star, 3 phase 50 Hz transformer has an iron loss of 1600 W and maximum efficiency occurs at 75% of full load. Determine maximum efficiency and efficiency on full load at 0.8 pf lagging.

   [Ans: 96.6%, 95.6%]

5. A 6.6 kV, 3 phase, Delta/star connection transformer has equivalent resistance of 1% and equivalent leakage reactance of 6%. The primary is supplied at 6.6 kV. Determine the ratio of voltages for delivering full load current at 4.8 kV, 0.8 pf lagging.

   [Ans: 0.437]

6. A 3-phase Delta/Star connected transformer rated at 2000 kVA, 6.6/0.4 kV gives the following test results.

S C Test	H V side:	400 V, 175 A and 17 kW
OC Test	LV side:	400 V, 150 A and 15 kW

   Determine the % resistance drop and % leakage reactance drop, % efficiency and % regulation on full load 0.8 pf leading.

   [Ans: 0.86%, 3.4% - 1.34%, 98%]

7. Three single-phase transformers connected in Delta/Delta, supply 300 A as the balanced load. One unit develops fault and is removed while two remaining transformers connected in open-Delta continue to supply the load. Determine the load current for same temperature rise.

   [Ans: 173.2 A]

8. From a 3-phase, 600 V Scott connected transformer system two single phase furnaces, I and II are supplied at 100 V. The furnace I is supplied from the Teaser transformer. Determine the line current on the three-phase side when (a) each furnace takes 600 kW at 0.8 pf lagging (b) furnace I takes 500 kW at unity pf and furnace II take 600 kW at 0.8 pf lagging.

   [Ans: (a) 143.8 A (b) 95.8 A]

9. A 3-phase, three windings star/star Delta connected transformer on open circuit takes 50 kVA at 0.15 pf lagging. Determine the Primary input kVA and power factor when for balanced loads the secondary delivers 870 A at 0.8 pf lagging and the tertiary delivers 260 line amperes at unity pf. Neglect the leakage impedances.

   [Ans: 1100 kVA, 0.8 pf]

10. Two 1000 kVA transformers having equal turns ratio are connected in parallel. The full load resistance drop is 1% and 6% for one transformer and 1.5% and 5% for the other transformer. A load 1600 kW at 0.8 pf is shared. Find the loads shared by them and the pf at which each transformer is operating.

    [Ans: 700 kW, 0.76 lag; 900 kW, 0.84 lag]

11. A 3-Phase balanced load of 50 kW at 340 V and 0.8 pf lagging is applied to a 3-phase Y-connected autotransformer. Find the currents in the input supply lines and output load lines. Also find the currents in the windings when supply voltage on the primary side is 400 V.

[Ans: 90 A, 106 A, 90 A, 16 A]

12. Three 1100/110 V, single-phase transformers are connected in Delta/Delta to supply 100 kW load at unity p.f. One of the transformer developed fault and is removed. What will be the currents following in each transformer when (a) all the three transformers operate, (b) two transformers operate?

[Ans: 303 A, 30.3 A, 525 A, 52.5 A]

CHAPTER 4
DC GENERATORS

4.1 Introduction to DC Machines

DC machines are the rotating machines which are available in a wide range of capacity. DC machines are available starting from a fractional HP to hundreds of HP. Small DC motors are used in watch, car-wiper, car starter etc. while the big machines find applications in rolling mills, cranes, traction etc.

A DC machine can be a DC generator or a DC motor. The DC generator converts mechanical energy into DC electrical energy while a DC motor converts DC electrical energy into mechanical energy. A DC machine consists of a stator body having salient poles projecting inwards, and a cylindrical rotor body. The salient poles act as electromagnets when a concentric field winding, placed on it, is energized by a DC supply. The distributed armature winding is embedded in the slots cut on the surface of the cylindrical rotor and it carries AC induced current.

4.2 Construction of DC Machines

There are two major parts in DC machines. One is the stator and the other is the rotor. The stator has a salient pole construction over which the DC concentric Field winding is placed. This field winding carries DC current through it. The sense of winding is such that the alternate poles become North and South poles. As a result of this, the main magnetic flux becomes available in the air gap (between the rotor and the stator) to interact with the armature conductors. The yoke of the stator is made up of Cold Rolled Grain Oriented (CRGO) steel. It is covered by a cast iron frame that has flanges. The yoke provides path for main Field flux. The machine is grouted on machine bed (cemented floor or platform) using nuts and bolts as shown in Fig. 4.1.

The rotor is also made up of CRGO steel sheets. It is a laminated solid cylindrical body which is mounted on a shaft with the help of wedge shaped shaft-key. The rotor contains an AC distributed Armature copper winding spreaded over slots on its outer periphery in semi-closed or closed manner.

The rotor also contains Commutator made up of copper segments which are connected with the armature winding. Stationary spring mounted Carbon Brushes are pressed over rotating commutator segments to bring the armature winding terminals out from the rotating body.

4.3 Principle of Operation of DC Machines

In case of a DC generator, the armature winding is made to move in the presence of main field, flux produced by field winding placed on stator. Additive emfs get induced in the armature conductors according to the Fleming's Right Hand (FRH) rule. Since the winding is closed, the current circulates in the armature winding. The nature of emf in individual conductor is alternating in nature, however, the winding layout and the brush-commutator arrangement are such that a DC emf output is obtained through the brush terminals.

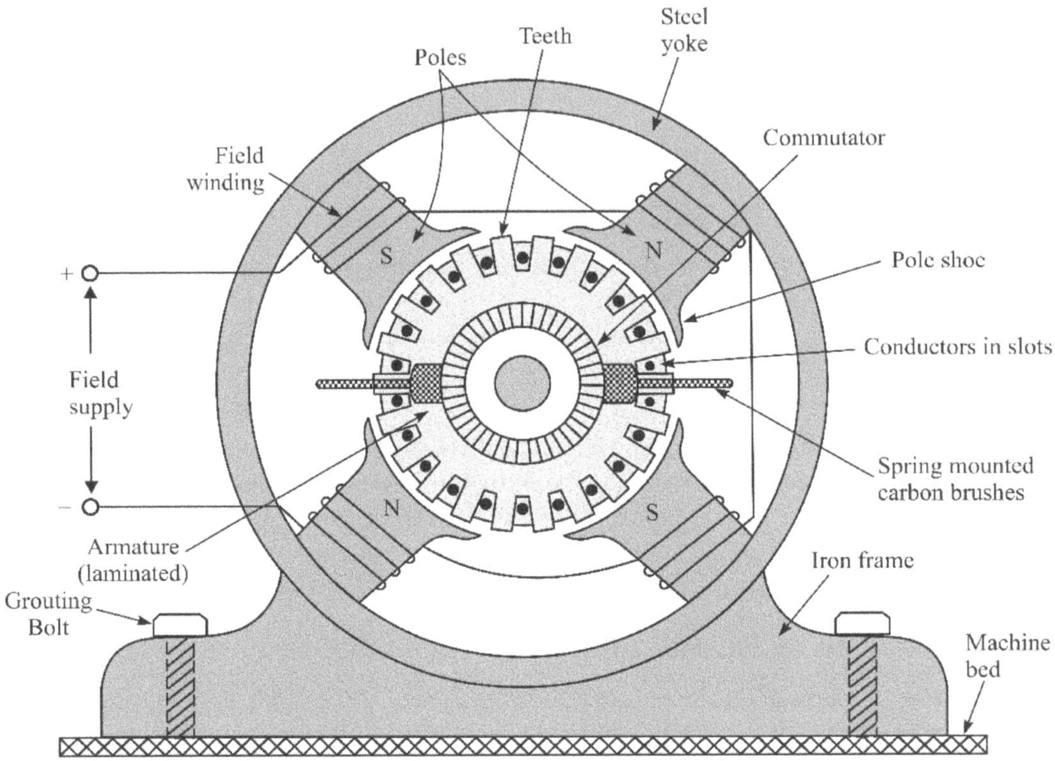

Figure 4.1 Construction of DC Machine

In case of a DC motor, the field winding is energized as usual to create the magnetic field in the air gap. A DC supply is fed to the armature winding through brush-commutator arrangement. The armature conductors, carrying current in the presence of the magnetic field, experience force according to Fleming's Left Hand (FLH) rule. The whole arrangement is such that the forces experienced by armature conductors lead to development of torque that makes the rotor rotate.

The brush – commutator arrangement acts as a converter to convert ac to dc in case of generator. DC machines work on the principle of electromechanical energy conversion.

4.4 Types of DC Machines

Energizing the field winding by applying DC voltage to it is known as excitation. The field winding produces magnetic field in the machine. The classification of DC machines is mainly based on the excitation method used. Broadly there are two ways of excitation. One, in which the field and the armature windings are separate, is termed as Separately Excited Machine. The other, in which the two windings are interconnected, is termed as Self Excited Machine. The self-excited machines are further classified based on the manner in which the armature winding is electrically connected to the field winding. The classification chart is given in Fig. 4.2.

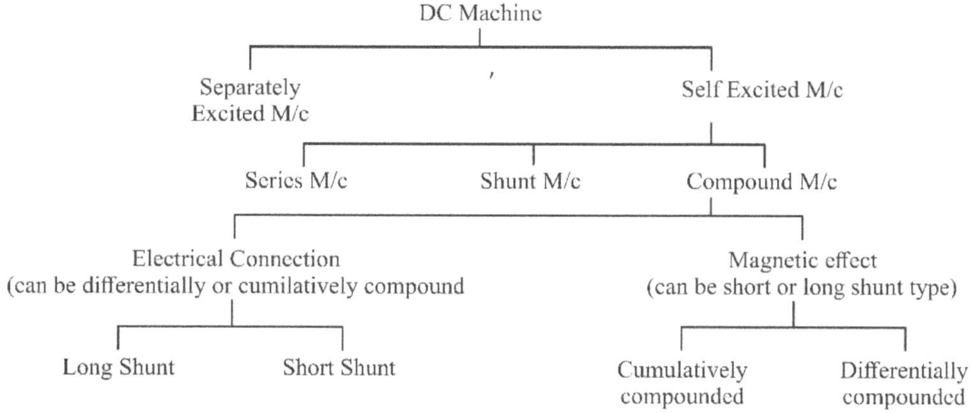

Figure 4.2 Classification of DC machines

4.5 Circuit Models

The electrical symbols of the field and armature circuits of DC machines are as follows Fig. 4.8.

(a) Field winding (b) Armature winding

Figure 4.3 Circuit Model

Using above the generating and the motoring modes of DC machines can be explained as follows.

4.5.1 Generating Mode

The rotor rotates in direction shown by speed 'ω'. The prime mover rotates the shaft and provides mechanical input to the machine and shown by 'T_m'. The electrical load connected across the armature terminals exerts electrical load torque 'Te' in opposition to the driving torque 'T_m'. The emf induced in armature of generator is expressed by 'E_a'. The voltage equation is given by

$$E_a = V_t + I_aR_a \text{ or } V_t = E_a - I_a R_a$$

Note that the direction of E_a and I_a are the same.

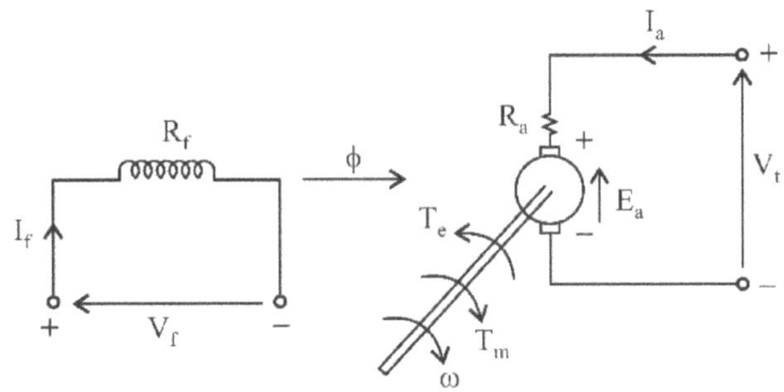

Figure 4.4 Generating mode of operation

4.5.2 Motoring Mode

The current drawn by armature winding develops electromagnetic torque 'T_e' due to which the rotor rotates in the same direction as shown by 'ω'. The mechanical load applied exerts the torque 'T_m' in direction opposite to that of 'T_e'. The emf developed in armature winding due to Lanze's law opposes the applied voltage V_t . The voltage equation is given by

$$E_b = V_t - I_aR_a \text{ or } V_t = E_b + I_aR_a$$

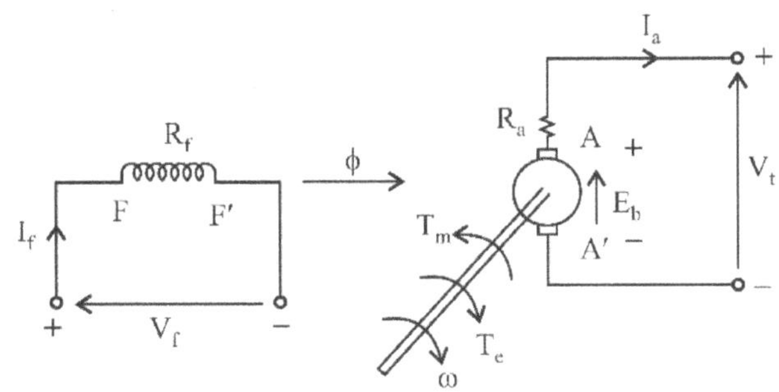

Figure 4.5 Motoring mode of operation

Note that the direction of E_b opposes the current I_a.

4.6 Methods of Excitation

The field winding of a DC machine can be excited either by an external source exclusively (called separately excited machines) or by an arrangement where no separate source is used (called self-exerted machines). The discussion in previous sections describes separately excited machines.

4.6.1 Circuit Models of various types of DC Machines

The circuit models of various types of DC machines are as follows.

A. DC separately excited Machine: Discussed in section 4.5.

B. Self-excited Machines:

B.1 DC Series Machine

DC Series Generator

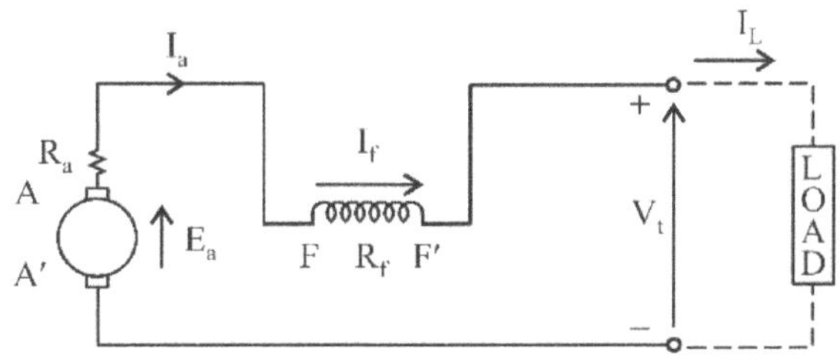

Figure 4.6(a) DC series generator

$I_L = I_f = I_a$

$V_t = E_a - I_a (R_a + R_f)$

DC Series Motor

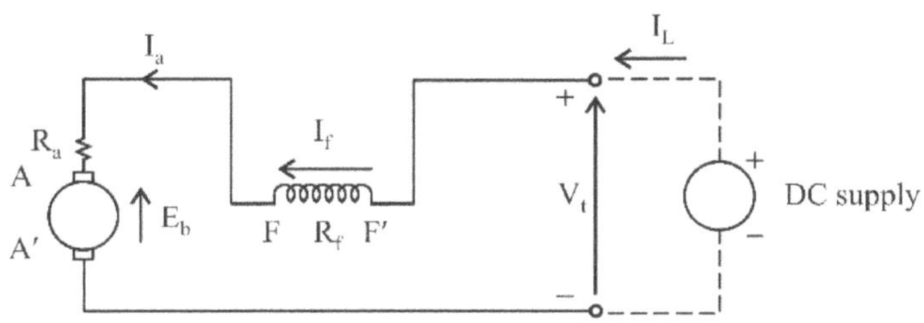

Figure 4.6(b) DC series motor

$I_a = I_f = I_L$

$V_t = E_b + I_a (R_a + R_f)$

B.2. DC Shunt Machine

DC Shunt Generator

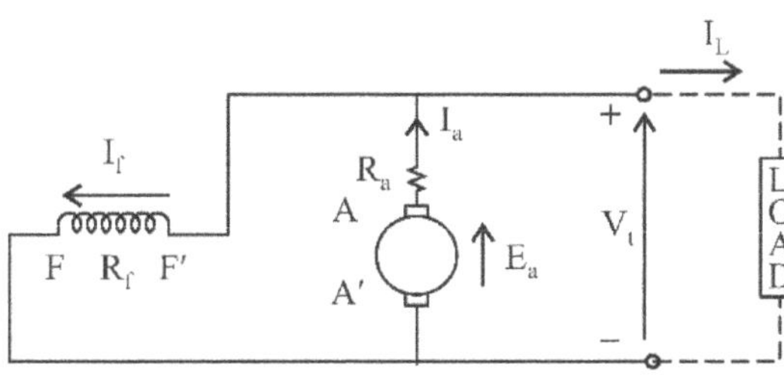

Figure 4.6(c) DC shunt generator

$I_L = I_a - I_f$

$V_t = I_f R_f = E_a - I_a R_a$

DC Shunt Motor

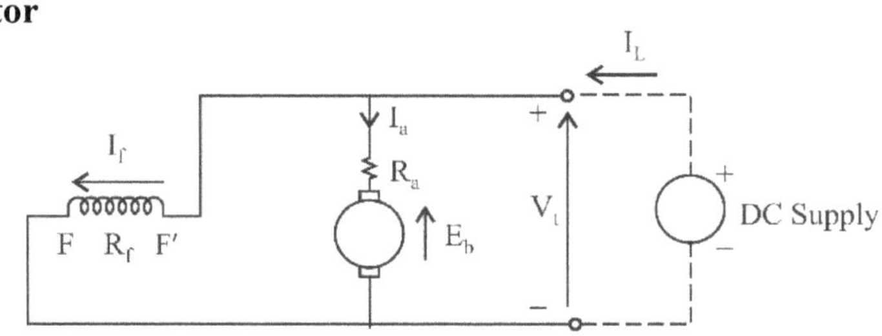

Figure 4.6(d) DC shunt motor

$I_L = I_a + I_f$

$V_t = I_f R_f = E_b + I_a R_a$

B.3 DC Compound Machine: In all compound machines, the shunt winding acts as the main field winding and the series winding as the auxiliary field winding. The following cases show 'Cumulative' compounding as both the components of field are additive (in the same direction) as shown. The Differential compounding, however can be obtained by reversing the shunt field winding terminals.

(a) DC Long Shunt Compound Generator

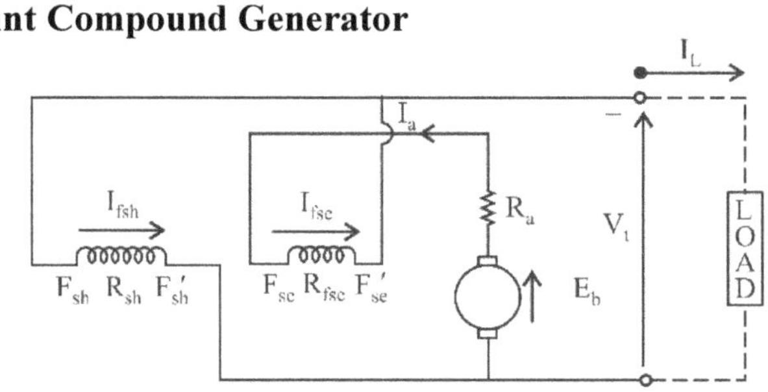

Figure 4.6(e) DC long shunt compound generator

$I_L = I_a - I_{fsh}$; $I_a = If_{se}$

$V_t = I_{fsh} R_{fsh} = E_a - I_a (R_a + R_{fse})$

(b) DC Long Shunt Compound Motor

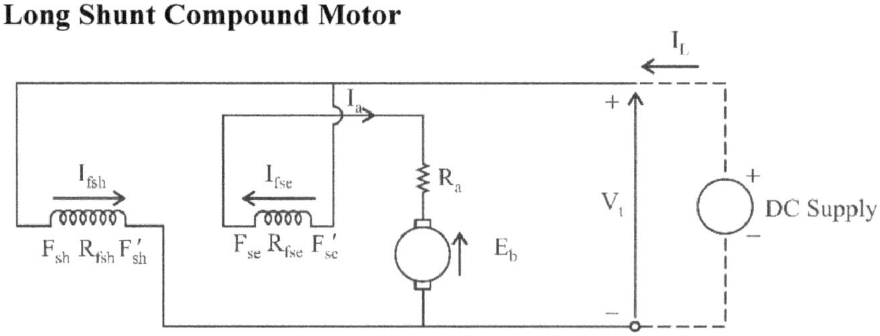

Figure 4.6(f) DC long shunt compound motor

$I_L = I_a + I_{sh}$; $I_a = I_{fse}$

$V_t = E_b + I_a (R_a + R_{fse}) = I_{fsh} R_{fsh}$

(c) DC Short Shunt Compound Generator

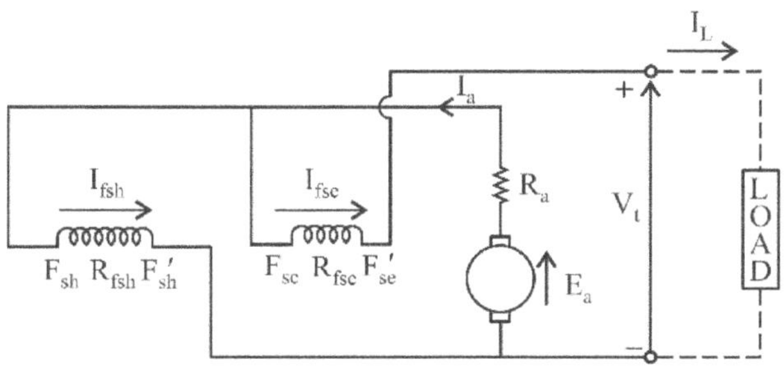

Figure 4.6(g) DC short shunt compound generator

$I_L = I_{fse}$; $I_a = I_L + I_{fsh}$;

$V_t = E_a - I_a (R_a + R_{fse}) = I_{fsh} R_{fsh} - I_{fse} R_{fse}$

(d) DC Short Shunt Compound Motor

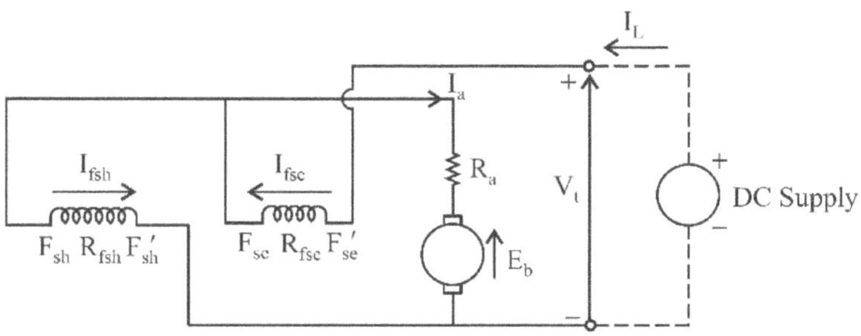

Figure 4.6(h) DC short shunt compound motor

$I_L = I_{fse}$; $I_a = I_L - I_{fsh}$;

$V_t = E_b + I_a (R_a + R_{fse}) = I_{fh} R_{fsh} + I_{fse} R_{fse}$

4.7 Armature Winding of DC Machines

The longitudinally laid down conductors in the slots cut on outer periphery of the armature on rotor, are distributed all around in a specific manner. The "Go" and corresponding 'Return' conductors fall about a pole pitch apart. The connections differ for the two types of armature windings, called Lap winding and Wave winding. The number of conductors placed in a slot are equal to or more than 02. In a winding diagram, the top conductor is shown by a solid line while the bottom one in the same slot is shown by a dotted line. The arrangement of winding will be explained with the help of winding table, ring diagram and developed winding diagram for specific cases for the lap and wave windings in the text to follow. Before that the various pitches are described.

Pole pitch (y): It is the number of conductors per pole

$$y = \frac{Z}{P}$$

where

Z = total number of conductors

P = number of poles

Back pitch (y_b): It is the number of conductors falling between a 'Go' and the corresponding 'Return' conductor.

$$y_b = y \pm k$$

where k = smallest number to make y_b odd in numbers.

Front pitch (y_f): It is the number of conductors falling between a 'Return' conductor to the 'Go' conductor of the next consecutive coil.

$$y_f = y_b - 2; \text{ for progressive winding}$$
$$= y_b + 2; \text{ for retrogressive winding}$$

Resultant or winding pitch (Y): It is the number of conductors falling between the 'Go' conductors of two consecutive coils.

$$Y = y_b - y_f$$
$$= 2 \quad \text{for progressive winding}$$
$$= -2 \text{ for} \quad \text{for retrogressive winding}$$

Commutator pitch (y_c): It is the number of commutator segments falling between the commutator segments connected to the 'Go' and the 'Return' conductors of one coil.

$y_c = 1$ for progressive lap winding,

$y_c = -1$ for retrogressive lap windings and

$y_c = y_b + y_f$ for wave winding.

Winding table: It is a table describing the complete layout of a winding.

Ring (sequence) diagram: It is a pictorial representation of the winding table.

Developed winding diagram: It is the actual layout of a winding. If a rotor is kept vertically and its surface is cut opened and flattened; the view thus developed is called the developed winding diagram.

The Fig. 4.7 shows the various pitches in Lap and Wave windings.

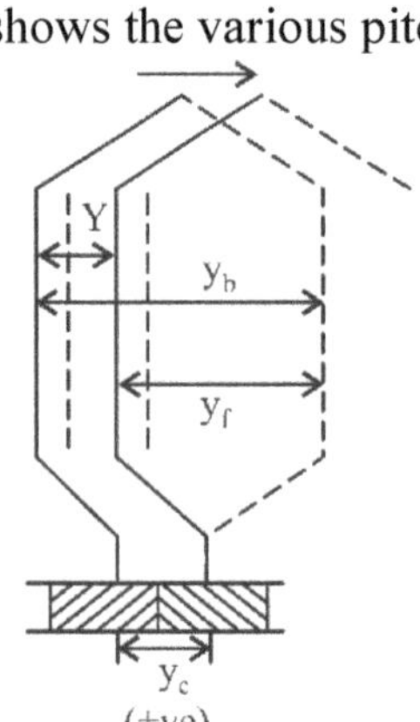

(a) Progressive Lap Winding

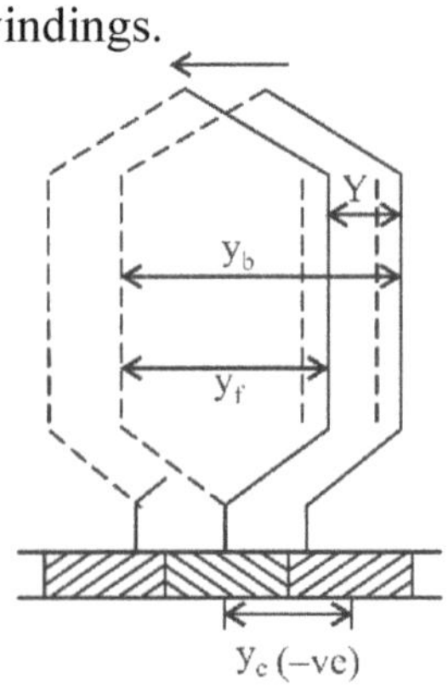

(b) Retrogressige Lap Winding

Wave Winding

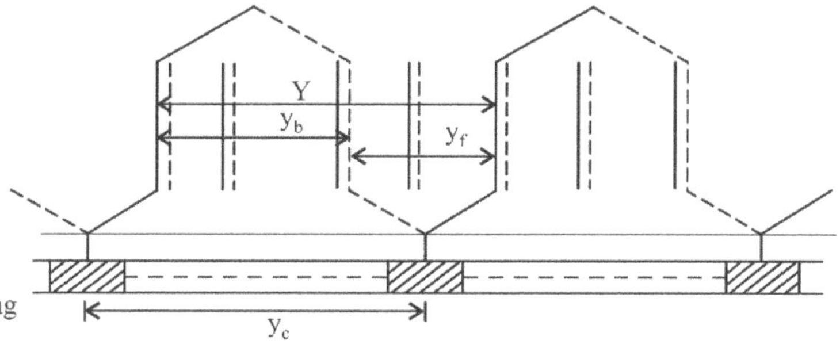

(c) Wave Winding

Figure 4.7 Various pitches

Example 1: The armature of a 4 pole DC machine has 25 slots. Two conductors per slot are arranged using lap winding. Find various pitches, develop the winding table, draw the ring diagram and developed winding diagram.

Solution: Given, P = 4, S = 25, 2 conductor/slot

Total number of conductors = 25 × 2 = 50

Pole pitch (y) $\frac{50}{4} = 12.5$

Hence $y_b = 13$ and $y_f = 11$

Winding Table of Ex 1

From	**To (+13) →**	**To (-11) →**	**To (+13) →**	**To (-11) →**	**To (+13) →**	**To (-11) →**
1	14	3	16	5	18	7
7	20	9	22	11	24	13
13	26	15	28	17	30	19
19	32	21	34	23	36	25
25	38	27	40	29	42	31
32	44	34	46	36	48	37
37	50	39	2	41	4	43
43	6	45	8	47	10	49
49	12	1				

The above table indicates that the winding gets closed.

Example 2: Develop the winding table and draw ring diagram and developed winding diagram for a 4 pole 25 plot wave wound DC machine. Two conductors per slots are kept.

Solution: Given P = 4 S = 25; $y = \frac{2 \times 25}{4} = 12.5$

Hence $y_b + y_f = 13$

Winding Table of Ex 2

From	To (+13) →	To (+13) →	To (+13) →	To
1	14	27	40	3
3	16	29	42	5
5	18	31	44	7
7	20	33	46	9
9	22	35	48	11
11	24	37	50	13
13	26	39	2	15
15	28	41	4	17
17	30	43	6	19
19	32	45	8	21
21	34	47	10	23
23	36	49	12	25
25	38	1		

Hence the winding gets closed.

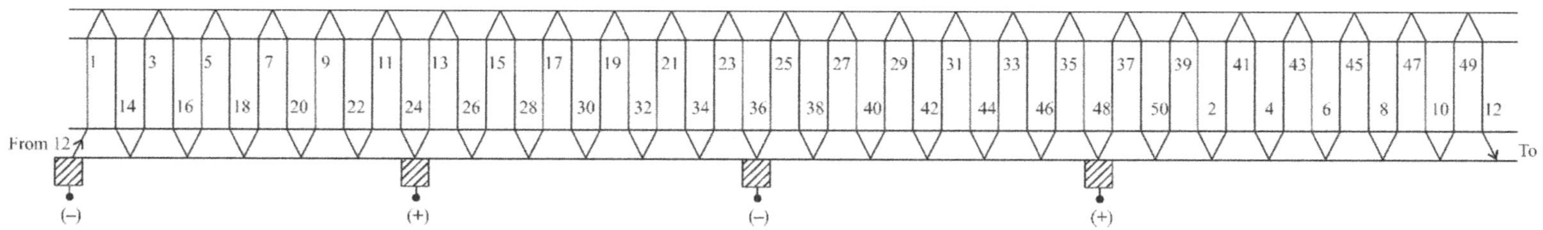

Figure 4.8(a) Ring diagram of Ex.1

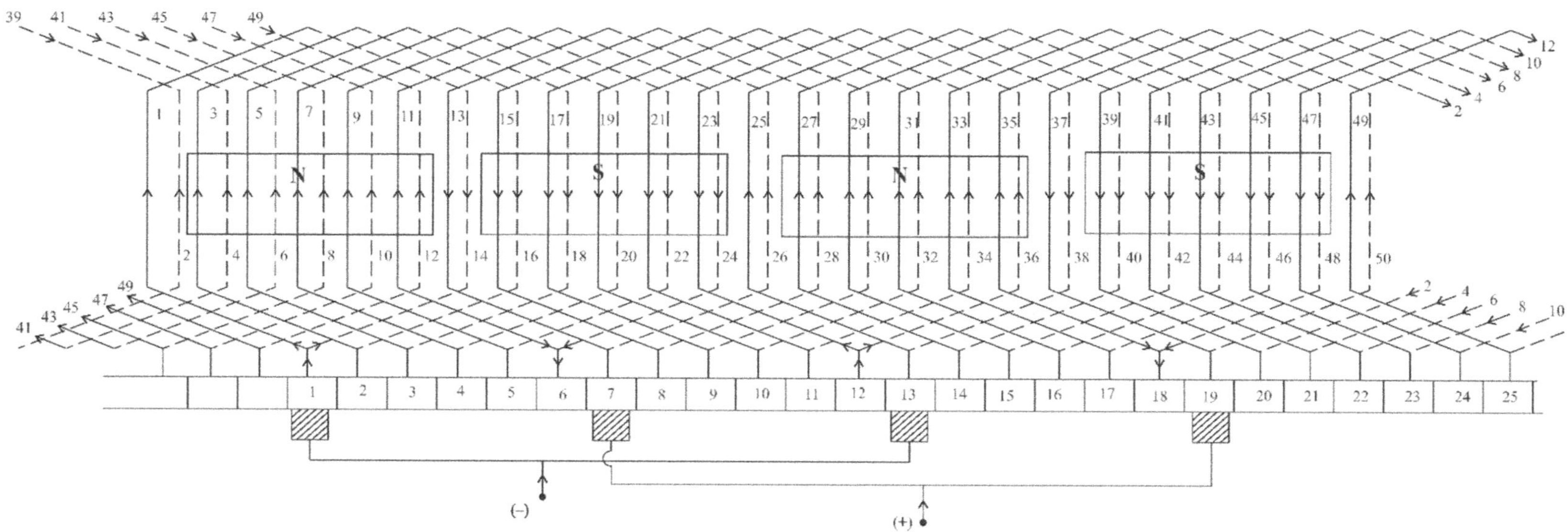

Figure 4.8(b) Developed winding Diagram of Ex.1

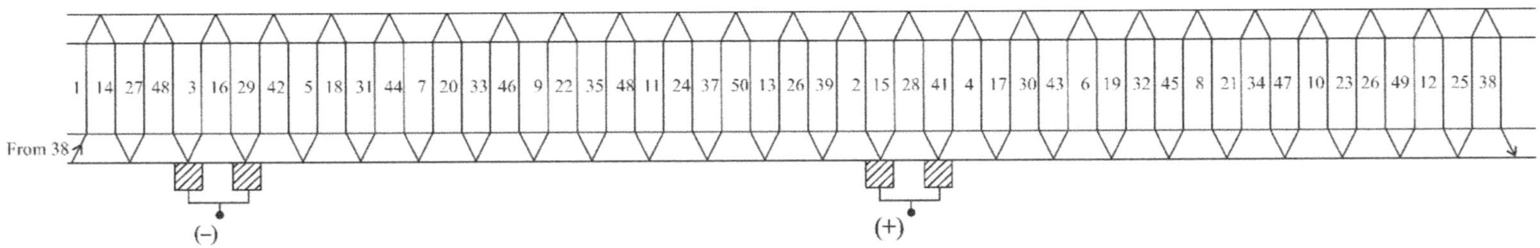

Figure 4.9(a) Ring diagram of Ex.1

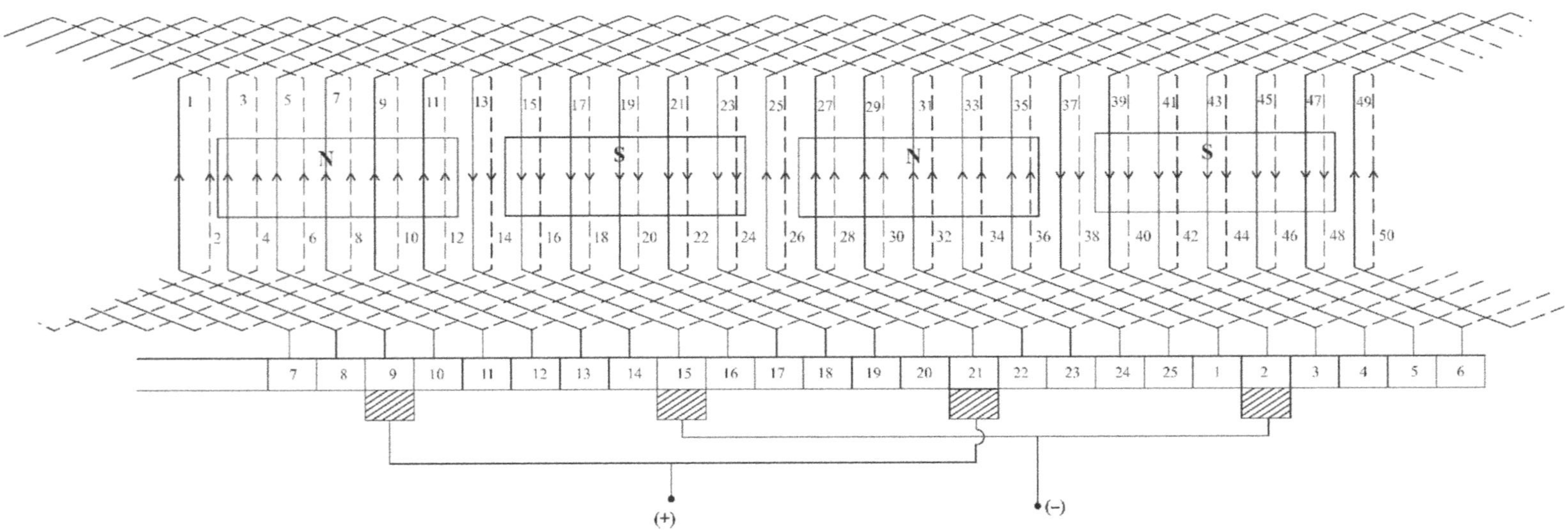

Figure 4.9(b) Developed winding Diagram of Ex.2

Example 3: A 4 pole DC machine has simpler lap winding of 120 coils arranged in 40 slots. Find out the condition such that the winding does not become a split winding.

Solution: Given, number of coils = 120, number of poles = 4

Hence the number of conductors = 120 × 2 = 240

$$\text{Pole pitch} = \frac{240}{4} = 60$$

Back pitch = 60 $\pm$ 1 = 61, 59

Also, it is seen that the number of conductors per slot $= \frac{240}{40} = 6$

Thus, the arrangements will look as shown in Fig.4.10 (a) and Fig. 4.10(b).

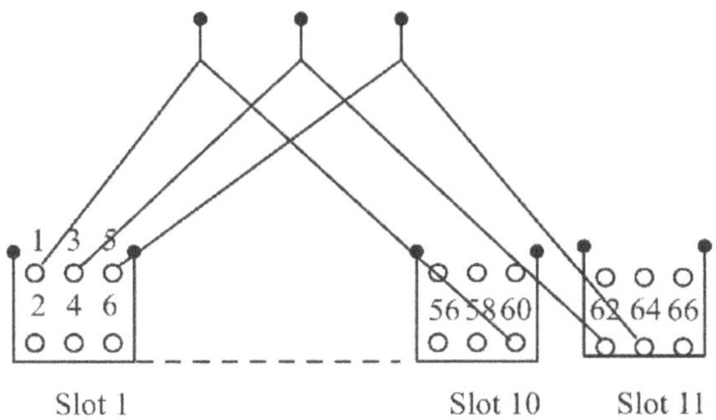

Figure 4.10(a) with yb = 59 (split winding)

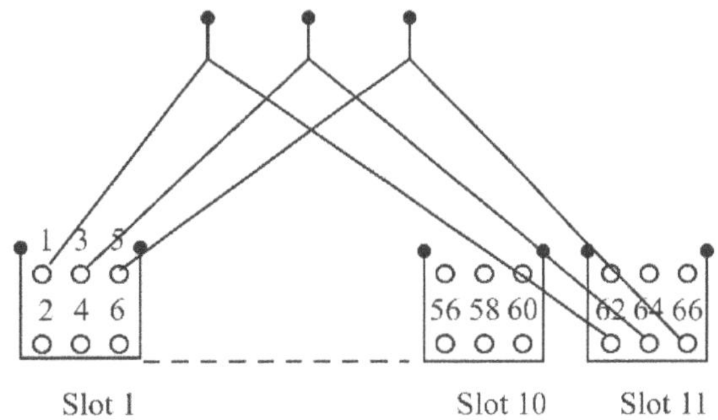

Figure 4.10(b) with yb = 61

From above example it is clear that the winding can be arranged without splitting only when $\frac{y_b - 1}{\text{Conductors per slot}}$ = Integer. Conductor per slot here is 6.

4.8 Equalizer Rings

When a DC machine operates, different emfs get induced depends in its armature coils spreaded over the rotor periphery. The amount of these emfs depends on the positioning of the coil under corresponding stator poles. Normally it is assumed that the coils occupying identical positions will develop the same emfs. However, it is difficult to obtain a perfect mechanical symmetry and hence the emfs may differ. This may lead to undesired copper loss due to circulating currents caused due to unequal emfs.

This problem is mitigated by placing equalizer rings at the overhang side of the armature. These are the short-circuited rings, to which the tappings from equipotential coils also connected. Thus, the circulating currents are prevented from flowing in the winding.

Distance between equipotential coils $= y_{eq} = \frac{2c}{P}$

Let 'm' be the number of equalizer rings to be placed.

(usually this number if from 10 to 20)

Hence, the total number of tappings = $\dfrac{mP}{2}$

Distance between adjacent tappings = $Y_{ph} = \dfrac{\text{No. of coils}}{\text{No. of taps}}$

or $Y_{ph} = \dfrac{2C}{mp} = \dfrac{z}{mp}$ (in terms of coils)

Example 4: Work out an arrangement for equalizer connections for an 8 pole simplex Lap wound DC machine having 200 coils. Take the number of rings equal to 10.

Solution:

$$Y_{eq} = \frac{z}{P} = \frac{2\times 200}{8} = 50 \text{ coils}$$

$$Y_{ph} = \frac{z}{mp} = \frac{2\times 200}{10\times 8} = 5 \text{ coils}$$

The layout of tappings to be connected to rings can be obtained by the table.

Ring No.	I	II	III	IV	V	VI	VII	VIII	IX	X
Coil Number	1	6	11	16	21	26	31	36	41	46
	51	56	61	66	71	76	81	86	91	96
	101	106	111	116	121	126	131	136	141	146
	151	156	161	166	171	176	181	186	191	196

4.9 Armature Reaction

In a DC machine the main magnetic field, responsible for its operation, is produced by the field winding. When a current flows through armature conductors, corresponding flux gets generated around them. The magnetic effect of this armature flux on the main field flux is called 'Armature Reaction'. The armature current depends on load current. Hence at no load condition there is no armature reaction. The armature reaction increases with increase in load current.

Consider a full pitch armature coil whose conductors lie at the centers of two adjacent poles. These conductors, when carry current produce their own magnetic fields surrounding them (right hand screw rule) which is proportional to the current carried by them. The net effect due to armature winding is to produce an mmf which is as shown in Fig.4.11.

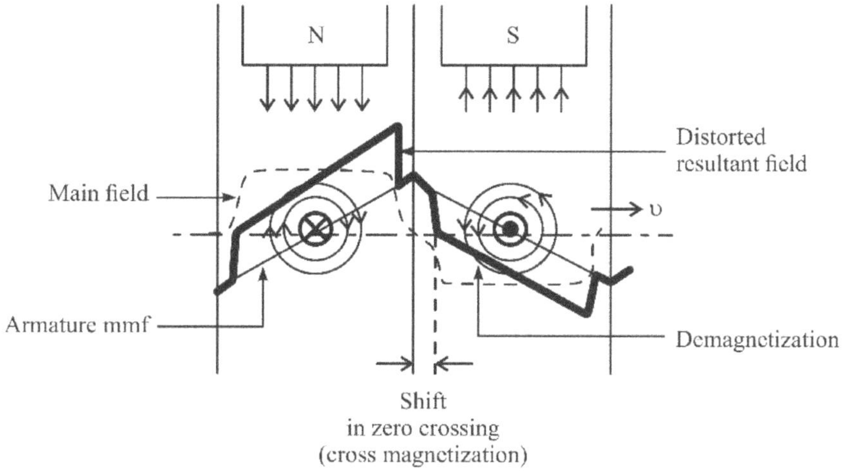

Figure 4.11 Distorted magnetic field

4.9.1 Shifting of Magnetic Neutral Plane (MNP)

The MNP gets shifted due to armature reaction. This is explained in Fig. 4.12.

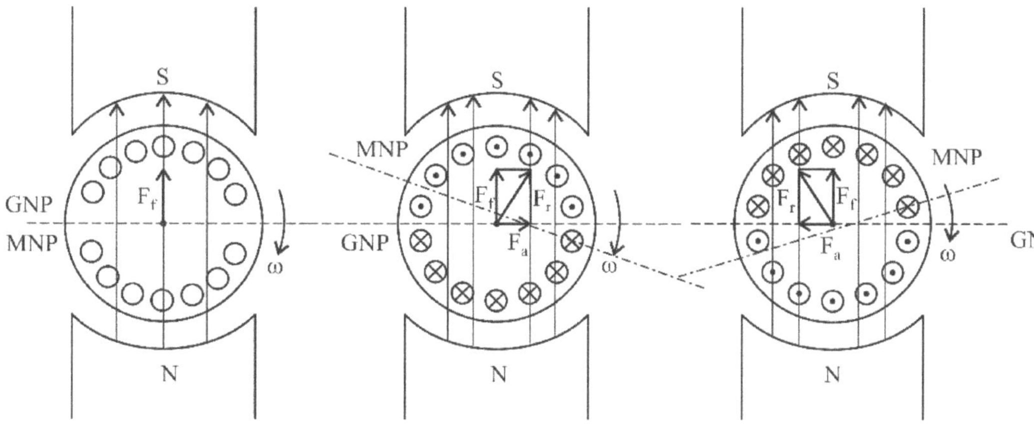

(a) With no armature current

(b) Generator (MNP shifts in the direction of rotation)

(c) Motor (MNP shifts opposite to direction of rotation)

Figure 4.12 Shift in MNP

Let the main field mnf be represented by F_f, the mmf due to armature current by F_a and the resultant of the two be given by Fr. In Fig.4.12 (a), if there is no current in the armature winding, the GNP (Geometric Neutral Plane) coincides with the MNP (Magnetic Neutral Plane).

Fig.4.12 (b) and Fig. 4.12 (c) show the conditions when the armature winding carries current in case of generator and motor respectively. The directions of rotation of armature is shown by 'ω'. Here one can find the direction of F_a by grabbing the armature with right hand keeping the direction of fingers pointing towards the current directions in armature conductors. The direction of thumb will then give the direction of F_a. The direction of F_f will remain unaltered in all three cases. In case of generator and motor the resultant emf 'f_r' is obtained as shown. The MNP always lies perpendicular to F_b. The diagrams explain that the MNP gets shifted in the direction of rotation in case of generators while the case is reverse in case of motors.

The resultant mnf can be as high as 1.3 times the mmf at no load (i.e mmf due to main field alone). Thus, the adverse effects due to armature reaction are:

(i) The main field gets demagnetized, reducing the induced emf.

(ii) The cross-magnetizing component of mmf due to armature reaction shifts the zero crossing of the emf waveform.

(iii) At the peaking portions of the emf wave form and the increased air gap flux density, cause greater iron loss in the teeth. This leads to increase in overall iron loss. The iron loss under loaded condition can be as high as 1.5 times to that at no load condition.

(iv) Problem of sparking and ring fire arises. Due to increased air gap flux density at places the potential difference between adjacent coils gets increased. If it becomes greater than 30 V, spark is formed. This may lead to formation of are ring around commutator.

(v) Delayed commutation is caused. Consider Fig. 4.13(a) where only main field mmf is present. Fig.4.13(b) shows a situation where only armature mmf is present.

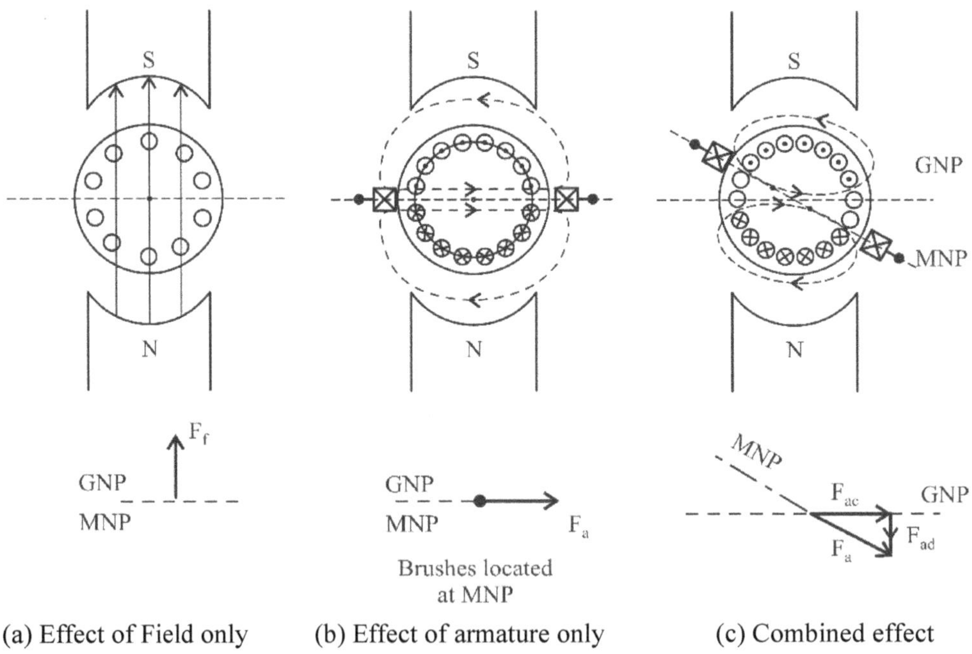

Figure 4.13 Brush location

When the two mmf (F_f and F_a) get super imposed, as shown in Fig.4.13(c) due to shift in MNP, the flux distribution becomes distorted. Its effect is shown by corresponding phasor diagrams. The armature mmf is resolved into two components called F_{ac} (Cross magnetizing component) and F_{ad} DC magnetizing component) of the armature mmf.

The Physical significance of armature reaction is explained in Fig.4.14.

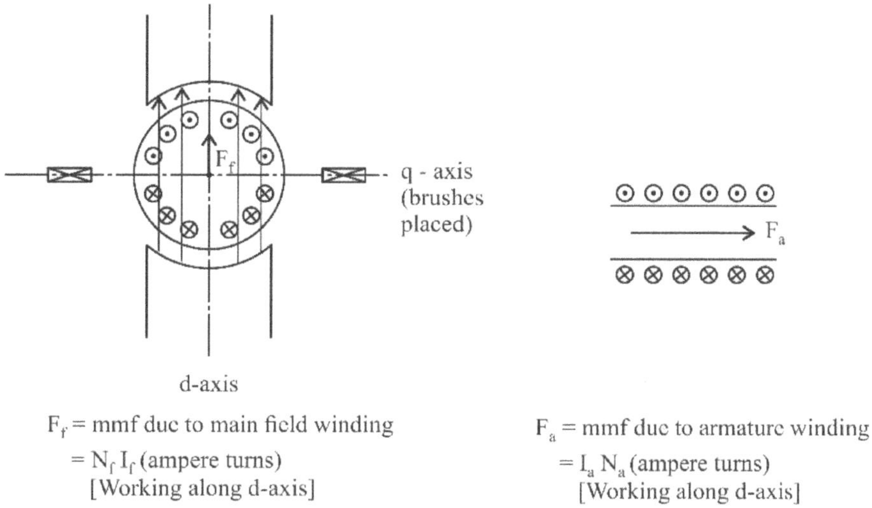

Figure 4.14 Mmf due to field and Armature winding

The developed diagram shown in Fig.4.15 elaborates the asymmetry caused due to armature reaction. As the direction of action of armature mmf (or flux) crosses the direction of action of main field mmf (or flux) perpendicularly, the phenomenon is called cross magnetization.

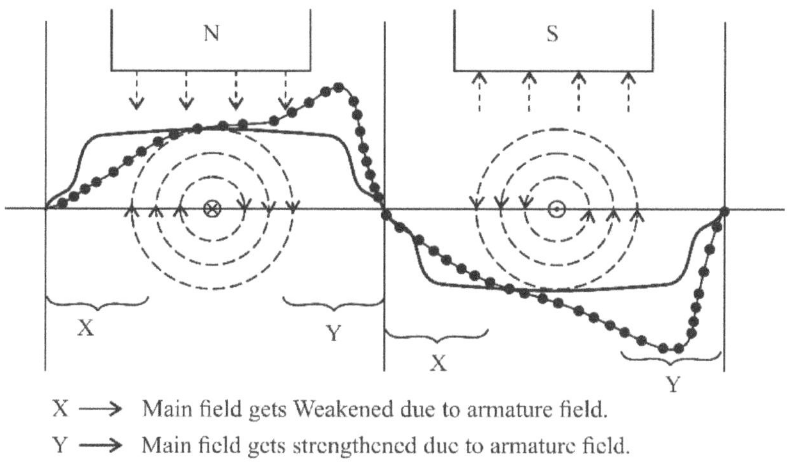

Figure 4.15 Developed diagram showing resultant mmf distribution

Due to symmetry, the weakening and strengthening of magnetic fields should be the same in amount. However, the magnetic saturation occurs in armature teeth. Hence the strengthening effect becomes less as compared to the weakening effect. The resultant flux, therefore decreases from the actual value. This is called demagnetization effect of armature reaction.

Fig. 4.15(a) is drawn to further explain the effect of armature reaction using two-solenoid concept.

When the brushes are shifted at MNP, the armature becomes equivalent to two-solenoids.

(a) AB-DC in the direction of GNP and

(b) AD – BC perpendicular to GNP.

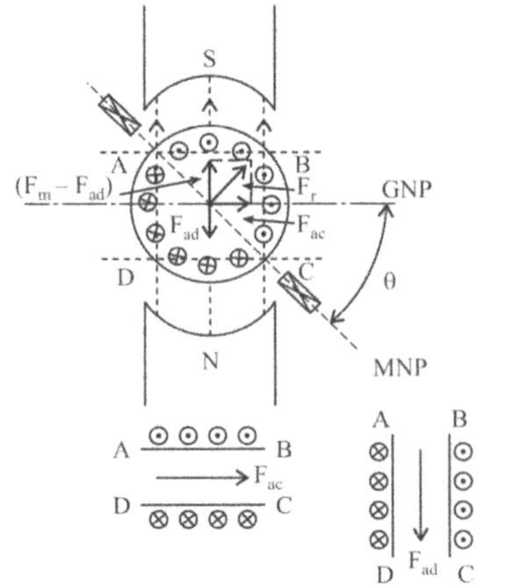

(a) Cross magnetizing component (b) Demagnetizing component

Figure 4.15(a) Two-solenoid concept

Let us now find out the values of F_{ad} and F_{ac}

I_a = Total armature current

Z = Total number of conductors

p = Number of pole pairs

a = Number of parallel paths

θ = Shift of brushes in mechanical degrees

Demagnetizing Mmf F_{ad}:

When the brushes are at GNP; the armature mmf per pole

$$F_a = AT_a = \frac{1}{2} \cdot \frac{Z}{2p} \cdot \frac{I_a}{a}$$

The demagnetizing effect is caused by the conductors of solenoid AD – BC. These conductors fall in the space covered by angle 4θ. This region is given by $\frac{4\theta}{360} r$

$$\text{Conductors per pole} = \frac{4\theta}{360} \cdot \frac{Z}{2p}$$

$$\therefore \qquad \text{Turns per pole} = \frac{1}{2} \times \frac{4\theta}{360} \cdot \frac{Z}{2p}$$

∴ Demagnetizing Amp turns per pole

$$A\,T_{ad} / \text{pole} = \frac{1}{4p} = \frac{4\theta}{360} Z = \frac{I_a}{a} = \left(\frac{4\theta}{360}\right) \frac{ZI_a}{4pa}$$

Cross magnetizing Mmf F_{ac}:

$$A\,T_{ac} / \text{pole} = \text{Total armature AT/pole} - AT_{ad} / \text{pole}$$

$$= \frac{ZI_a}{4pa} - \left(\frac{4\theta}{360}\right) - \frac{ZI_a}{4pa}$$

∴ Cross magnetizing Amp turn per pole

$$\frac{AT_{ac}}{\text{pole}} = \frac{ZI_a}{4pa}\left(1 - \frac{4\theta}{360}\right)$$

4.10 Mitigation of Adverse Effects of Armature Reaction

Armature reaction is an undesirable yet inevitable phenomenon. The amount of armature reaction and hence the shifting of brushes, is a function of armature (load) current. It is practically not possible to change the brush location with changing load. Hence use of interpoles or compensating winding is done to nullify the effect of armature reaction.

(a) **Interpoles:** Small auxiliary electromagnetic poles are placed in quadrature axis (at GNP) to nullify the cross magnetizing component of armature reaction.

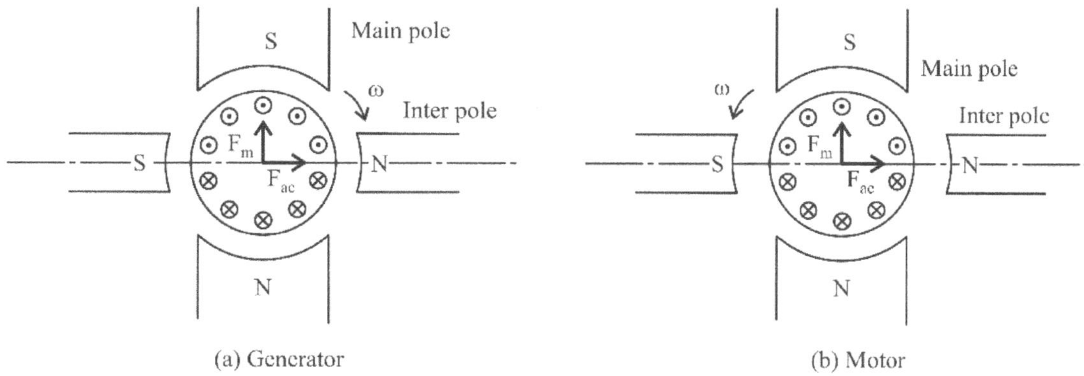

Figure 4.16 Interpoles

The armature conductors encounter opposite polarity interpole in the direction of rotation in generators whereas they encounter same polarity interpole in motor. The winding on interpoles is kept in series with the main field winding. It is clear from the Fig.4.16 that the cross-magnetizing component of armature reaction gets nullified by the magnetic field of the interpoles. Effect of cross magnetization can be completely removed by adequately deciding the number of turns of interpole winding as both F_{ac} and F_{ip} (interpole amp turn) are directly proportional to the armature (load) current.

Leakage coefficient: It is necessary that in order to nullify F_{ac} completely, the F_{ip} must be equal to or slightly greater than F_{ac}.

$$\text{Leakage coefficient} = \frac{\text{Total ampere turns on interpole}}{\text{actual } AT_c}$$

$$K_l = \frac{N_{ip} I_{ip}}{AT_c}$$

where

N_{ip} = Number of turns of interpole winding

I_{ip} = current carried by interpole winding = $I_a/2p$

Hence $$N_{ip} = \frac{K_l . AT_c}{I_{ip}};$$

where

$K_l = 1.1$ or 1.2

(b) Compensating Winding: The function of the compensating winding is to neutralize the effect of distortion of main field flux under pole-shoe region due to armature reaction. The compensating winding carries armature current to provide required compensation at all loads.

Let b_p = pole arc, y = pole pitch $= \dfrac{\pi D}{2p}$

Total armature mmf per pole $= \dfrac{Z}{4p}\dfrac{I_a}{a}$

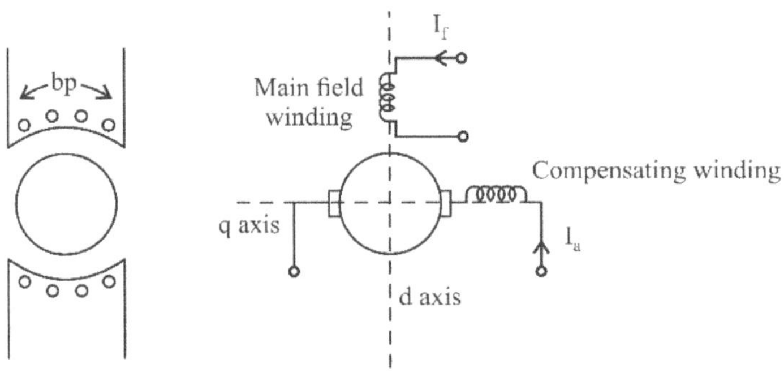

Figure 4.17 Compensating winding

The ampere turn of compensating winding is given by

$$\frac{N_c}{p} \times I_a = K_l \frac{Z}{4p} \cdot \frac{I_a}{a} \cdot \frac{b_p}{y}$$

$$\therefore \qquad N_c = K_l \frac{Z}{4a} \cdot \frac{b_p}{y}$$

where

N_c = Number of turns of compensating winding

4.11 Commutation

It is a process of converting alternating current of the armature coil into direct current in the external circuit, with the help of commutator and brush arrangement in a DC machine.

Let

I_c = armature conductor current

$I_b = 2I_c$ = current carried by brush

W_c = width of commutator segment

W_m = width of mica insulation between segments

W_b = width of brush = $(W_a + W_m)$

v_c = peripheral speed of commutator

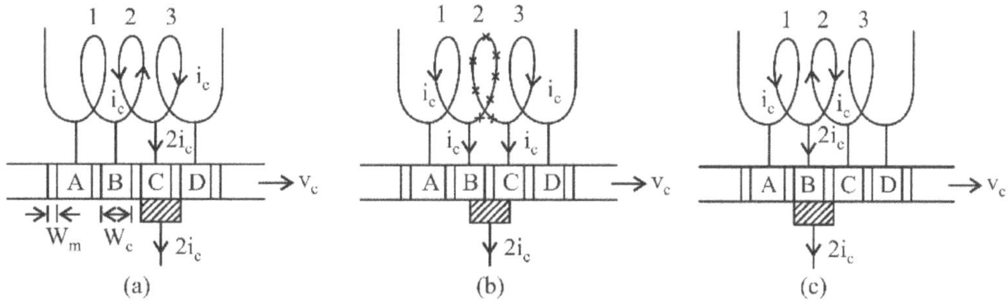

Figure 4.18 Commutation

Consider three coils 1, 2, 3 to understand the commutation process. The coils and the commutator segments are moving towards right with a peripheral velocity v_c. The stationary brush collects current from the commutator segments that come under its contact. The brush current is contributed by 2 coils at a time. Coils 2-3, coil 1-3 and coil 1-2 contribute when the brush contacts segment C, B-C and B respectively as shown in Fig.4.18(a), (b) and (c) respectively. A close observation of coil 2 reveals that the current through it is anticlockwise, zero and clockwise during commutation process. Thus, though the direction of current reverses in armature conductors the brushes always collect unidirectional current. The coil undergoing commutation gets short circuited during change of direction of current through it as can be seen in Fig. 4.18(b).

Period of Commutation: It is that time internal in which the brush contact shifts from one commutator segment to the next.

$$t_c = \frac{W_m + W_c}{v_c}$$

It is the time in which a coil current gets reversed from $+i_c$ to $-i_c$. Ideally, this change should be linear. However, there occurs some delay in commutation due to inductive effect of the coil undergoing commutation. Sudden current reversal in this coil induces emf $L_c \frac{di_c}{dt}$, which opposes the process of commutation and delays the same as shown in Fig. 4.19.

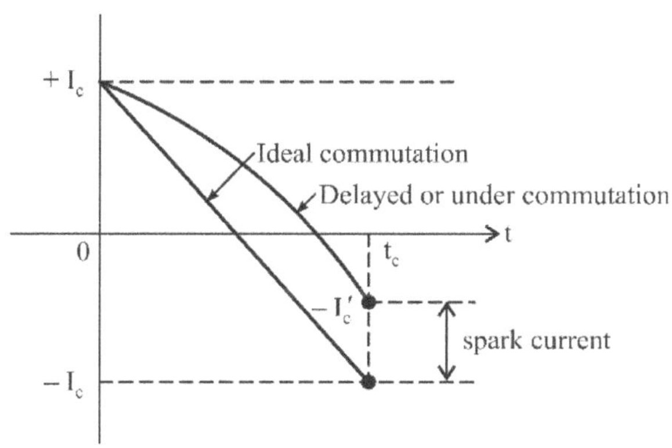

Figure 4.19 Delayed Commutation

Another reason for delayed commutation is the placement of brushes. It is not possible to shift the placement of brushes according to the shift occurring in MNP with the changing load. If the brushes are not placed exactly on MNP, the emf induced due to cross magnetizing component of armature reaction can cause under or over commutation depending on load.

A shift of commutation from its ideal characteristic results in sparking at brushes. This deteriorates the carbon brushes and commutator segments and reduces their lives. Resistance commutation or voltage commutation are used to overcome the problem of delayed commutation.

(a) **Resistance Commutation:** High contact resistance between the commutator segments and brushes are achieved by using carbon brushes. They add resistance to the circuit of commutating coil thereby reducing the time constant (L/R) of the current transient. This helps in faster change in current direction. Carbon brushes also help reduce commutation wear and are easily replaceable.

(b) **Voltage Commutation:** Interpoles are placed in q-axis which produce flux. This flux produces rotational voltage in the commutating coil in such a direction as to nullify the effect of reactance voltage. As the interpole winding is connected in series with the armature winding, the variation in cross magnetizing component of armature reaction is exactly nullified by the interpole strength. Thus, the variation of load does not lead to commutation problem

The airgap under interpoles is kept larger than that under main poles, so that their magnetic circuit is linear, resulting in cancellation of the reactance voltage (a linear derivative term) at all loads.

4.12 Characteristics of DC Generators

The performances of a DC generator is described by its various characteristics. Consider a DC generator whose rotor is rotated at a constant (rated) speed (ω rad/sec). The field winding is separately (or self) excited. An emf E_a gets induced in the armature winding which depends on the amount of excitation. R_a is the resistance of the armature winding. As shown in Fig.4.20, when there is no current through the armature winding, the voltage measured across armature terminals is same as induced emf E_a. Now when an electrical load R_L is applied across the armature terminals, current flows through the armature winding causing I_aR_a drop to appear. Thus, the voltage measured across armature terminals V_t. is less than the actual induced emf E_a.

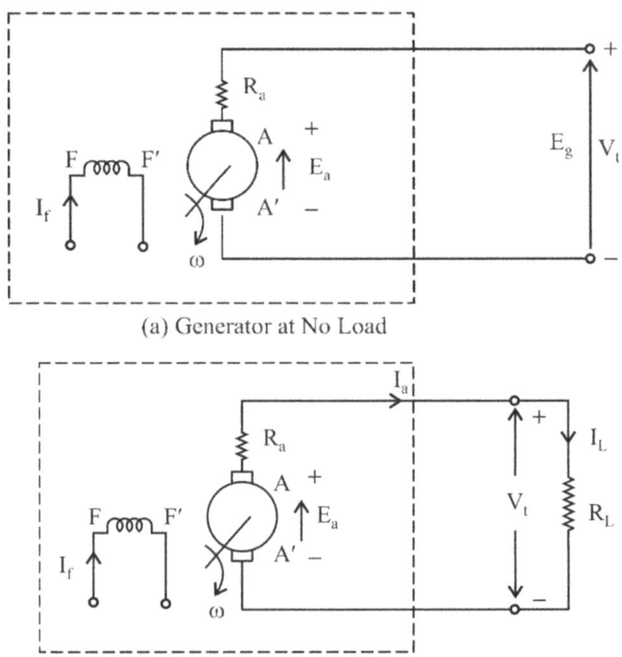

Fig.4.20 DC Generator

The circuit external to the generator is the load resistance R_L. The current drawn by it is I_L and the voltage drop across R_L is the terminal voltage V_t. Hence it is seen that induced emf E_a cannot be measured under loaded condition of the generator. What can be measured across armature terminal is Vt, which is equal to $(E_a - I_aR_a)$.

The following characteristics describe the performance of a DC generator.

1. Magnetizing or open circuit characteristic (OCC) [E_a Vs I_f]
2. External characteristic or Load characteristic [V_t Vs I_L]
3. Internal characteristic [E_a Vs I_a]
4. Regulation curve [I_a Vs I_f]

4.13 Open Circuit Characteristic of DC Generators

To obtain the OCC of DC generators of all types, the armature is run at a constant (rated) speed by the prime mover. The open circuit voltage is measured and plotted against the field excitation current I_f. If is gradually increased till 1.1 to 1.25 times the rated voltage is generated.

4.13.1 Separately Excited DC Generator

Fig.4.21 shows the circuit diagram and the OCC of separately excited DC generator. This machine gives a wide range of open circuit voltage V_{OC}.

We know that $E_a \propto \phi\ \omega$

Now if ω = constant,

$E_a \propto \phi$

Since $\phi \propto I_f$

Hence $Ea \propto I_f$

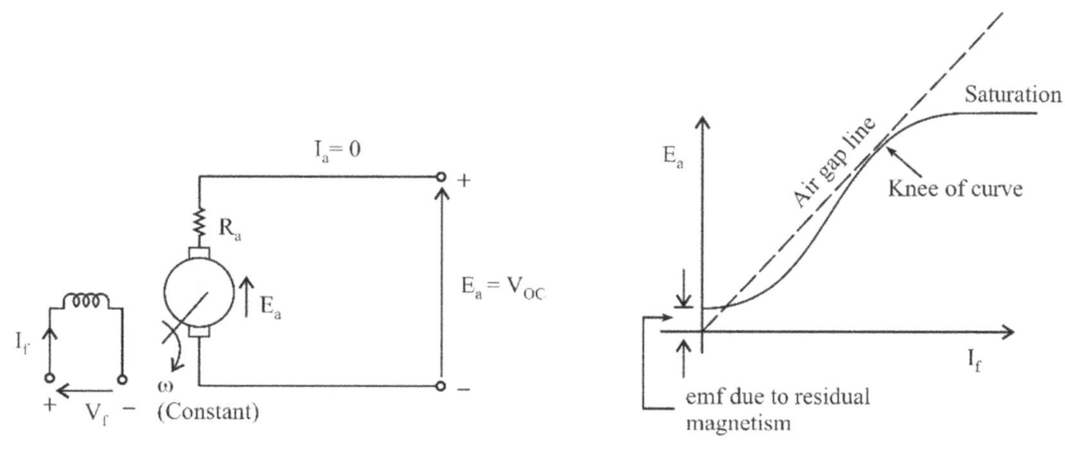

(a) Circuit (b) Characteristic

Figure 4.21 OCC of separately excited DC generator

The OCC is a plot of open circuit voltage (which is same as E_a) with the field excitation current I_f drawn by the generator at rated speed. There appears a small emf even when the excitation current is zero. It is because the machine would have been previously subjected to magnetization. The poles retain some residual magnetism. This residual magnetic is capable of generating the small emf as shown in Fig.4.21(b).

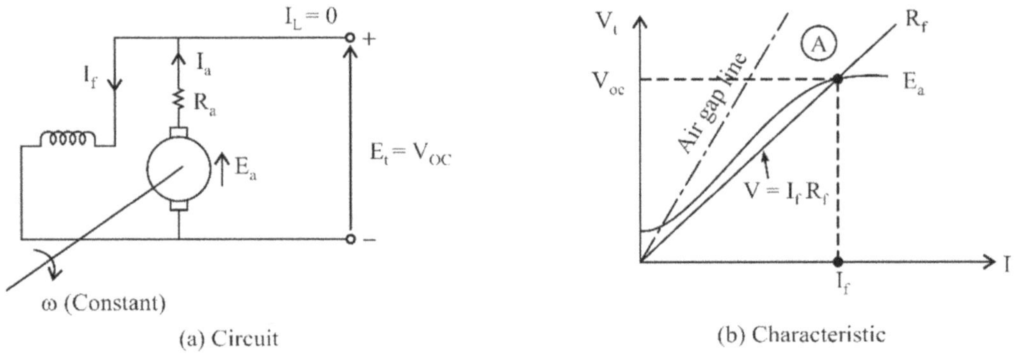

Figure 4.22 OCC of DC shunt generator

4.13.2 Self-Excited DC Shunt Generator

Refer to Fig.4.22(a) and assume that the field circuit is switched on when the armature has acquired the rated speed. A small emf gets reduced in the armature due to residual magnetism. This leads to circulation of a small current in the field circuit. The field poles get strengthened. As a consequence of this, more emf is generated by the armature which further enhances the field current. The process of development of emf keeps growing till the magnetic saturation of the poles occurs as shown in Fig. 4.22(b).

At point of operation 'A' the two conditions are fulfilled.

$V_{oc} = I_f R_f$ and $V_{oc} \approx E_a$

We know that

$I_a = I_f$; as $I_L = 0$ at no load

$E_a = V_t - I_a R_a$: If R_a is neglected, as it is very small, then

$E_a = V_t$;

where V_t is the measured open circuit voltage V_{oc}. Hence $V_{oc} = V_t$

In field circuit

$V_{OC} = I_f R_f$

Also, we know that $\phi \propto I_f$ and $E_a \propto \phi$, hence $E_a \propto I_f$

Thus, neglecting armature winding resistance

$E_a \approx V_{oc} \propto I_f$

The OCC is a plot of this relationship as shown in Fig.4.22 (b). Although it does not follow a linear relationship due to magnetic properties involved.

4.13.2.1 Building up of EMF

The process of building up of emf is shown in Fig.4.23. E_1 is the emf induced due to residual magnetism. This emf E_1, circulates current I_f in the field circuit. The value of I_f is decided by the field winding resistance line R_f. Now as the current I_f circulates in the field winding, the poles get strengthened to induce emf E_2. Amount of E_2 is decided by the OCC of the machine. E_2 causes current I_{f2} to flow in the field circuit, and the process of building up of emf keeps progressing.

$$E_1 \rightarrow I_{f_1} \rightarrow E_2 \rightarrow I_{f_2} \rightarrow E_3 \rightarrow I_{f_3} \rightarrow E_4 \rightarrow I_{f_4} \rightarrow E_5 \rightarrow I_{f_5} \rightarrow E_a \rightarrow I_f$$

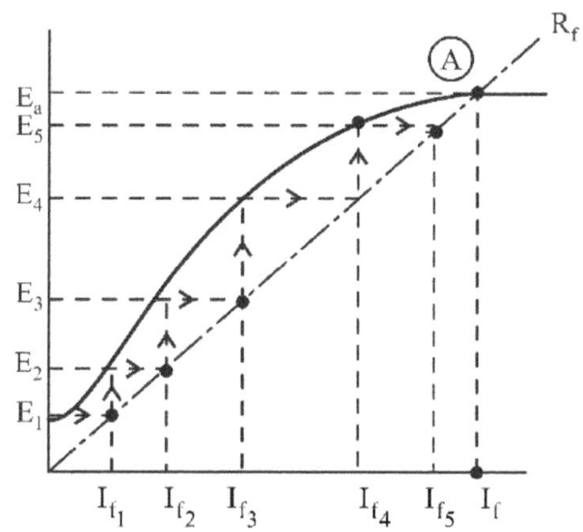

Figure 4.23 Building up of Emf

The point of operation 'A' is finally reached where the process stops. This is the operating point decided by the intersection of OCC with R_f line. At this point, both the characteristics, the OCC and the field resistance R_f, get satisfied and the machine keeps operating.

4.13.2.2 Critical Field Resistance

Refer to Fig. 4.24. Here the speed n is constant and the field resistance R_f is varied. Let the generator run at rated speed of n rpm. The point of operation would shift with the change in field resistance R_f.

If the value of R_f is increased from R_{f_1} to R_{f_2}; the value of open circuit voltage, available at generator terminals, would decrease from V_{oc_1} to V_{oc_2}. It can be seen that a further increase in field resistance to a value R_{f_3} no voltage would be available; as the R_{f_3} line does not intersect/touch the OCC. The machine does not get excited and the terminal voltage continues to be equal to that generated by residual magnetism only. Thus, the field resistance line tangential to the linear portion of OCC can be called the 'Critical field resistance' (R_{fc}). It is a must for satisfactory operation that $R_f \not> R_{fc}$. That is, the field resistance should not be greater than the critical field resistance.

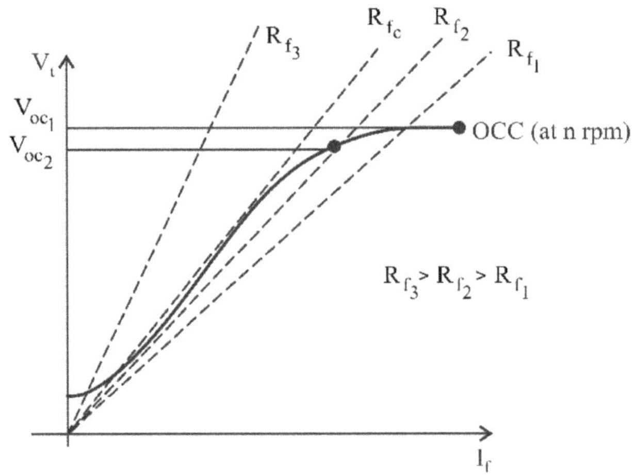

Figure 4.24 Critical field resistance

4.13.2.3 Critical Speed

Refer to Fig.4.25. Here the field resistance R_f is kept constant and the speed is varied. Let the generator have a field resistance equal to R_f. It generates voltage V_{oc_1} when it runs at n_1 rpm. Now if the speed is decreased to n_2; the generated voltage drops down to V_{oc_2}. At further lower speed of n_3, the OCC and the R_f line do not intersect each other and hence the generator would generate only the voltage corresponding to the residual magnetism. The OCC to which line R_f is tangential can be called as the 'Critical speed' (n_c), below which the generator does not get excited. Thus, in order to get a satisfactory operation of the generator $n > n_c$. That is, the speed of generator should always be greater than the critical speed.

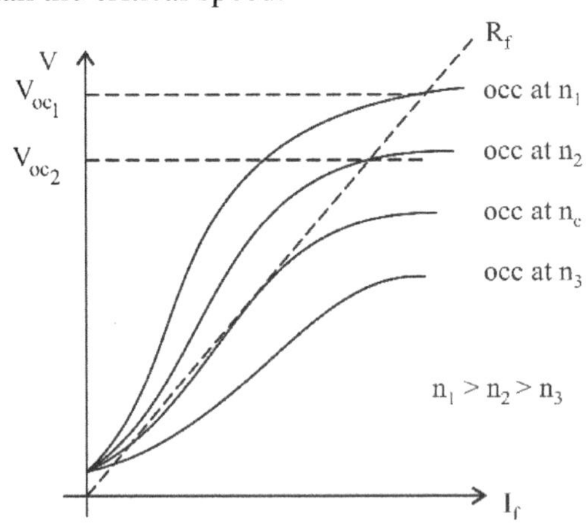

Figure 4.25 Critical speed

4.13.3 Self-Excited DC Series Generator

The armature and field windings of a series motor are connected in series with each other. To obtain the open circuit characteristics, it is required to increase the field current and get the corresponding rise in induced emf across armature winding. The armature current, which is same as field current and load current, can circulate only when a load is connected across the generator terminals. Hence, as such, it is not possible to obtain OCC of DC series generator. The OCC of the machine can only be obtained by operating it as a separately excited machine.

To summarize, the following conditions are required to be fulfilled in order to make a generator operate.

1. The poles of the machines must have residual magnetism.
2. Speed (n) $\geq$ critical speed (n_c).
3. Field resistance (R_f) $\leq$ critical field resistance (R_{fc}).
4. The polarity of field winding should support the residual magnetism. Else the residual magnetism would get nullified and the generator would not produce any emf.

4.14 External and Internal Characteristics of DC Generators

The external characteristic of DC generator is plotted between the load current and the terminal voltage. The terminal voltage drops with the increase in load current due to:

(i) Voltage drop across armature winding resistance
(ii) Voltage drop across interpole/compensating winding
(iii) Voltage drop across brushes
(iv) Voltage drop due to armature reaction.

The internal characteristic of DC generators is plotted between the load current and the induced emf. Out of the above four voltage drops mentioned, the first three vary linearly with the load current. The armature reaction, however, does not vary linearly with the load current.

It is not possible to plot internal characteristic directly. The internal characteristic is obtained by adding the drops to the external characteristics.

4.14.1 Separately Excited DC Generators

Fig.4.26 shows the circuit diagram and the External and Internal characteristics of a separately excited DC generator.

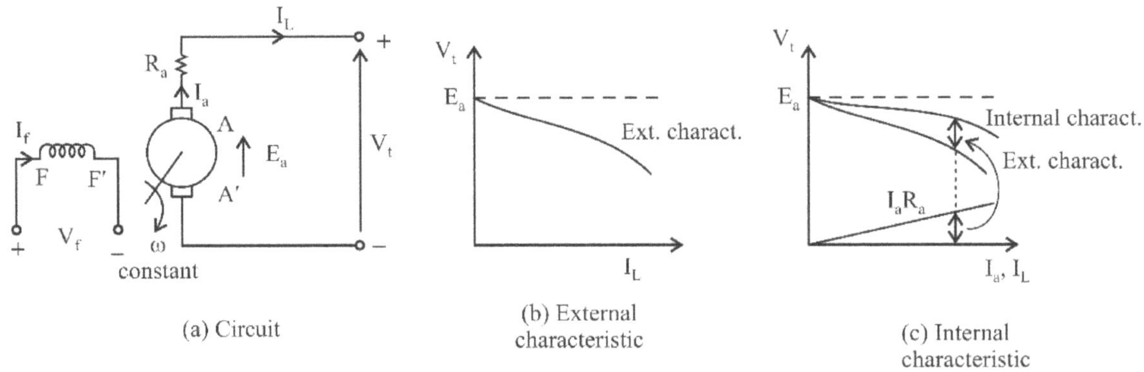

Figure 4.26 Internal and external characteristics of separately excited DC generator

The prime mover is run at a constant speed and the field current is also maintained constant. The load is gradually increased. It is observed that the terminal voltage drops with increase in load. Plot of terminal voltage against load current is known as external characteristics shown in Fig.4.26(b).

In order to obtain internal characteristic, the external characteristic is plotted as shown in Fig. 4.26(c). On current axis, same scale is used for I_L and I_a. The armature drop is plotted as I_aR_a. We know that $I_a = I_L$. For different values of I_L, the corresponding I_aR_a drop is added to the external characteristic. By doing this, the internal characteristic is obtained. Thus, the gap between the internal and the external characteristics represents the armature drop. The gap between constant V_t line and the internal characteristic represents the drop due to armature reaction and drops in brush, compensating/interpole windings. As the later drops are smaller as compared to armature reaction drop, it is normally considered to be caused by armature reaction only.

4.14.2 Self-Excited DC Shunt Generators

Fig.4.27 shows the circuit diagram and the external and internal characteristics of DC shunt generator.

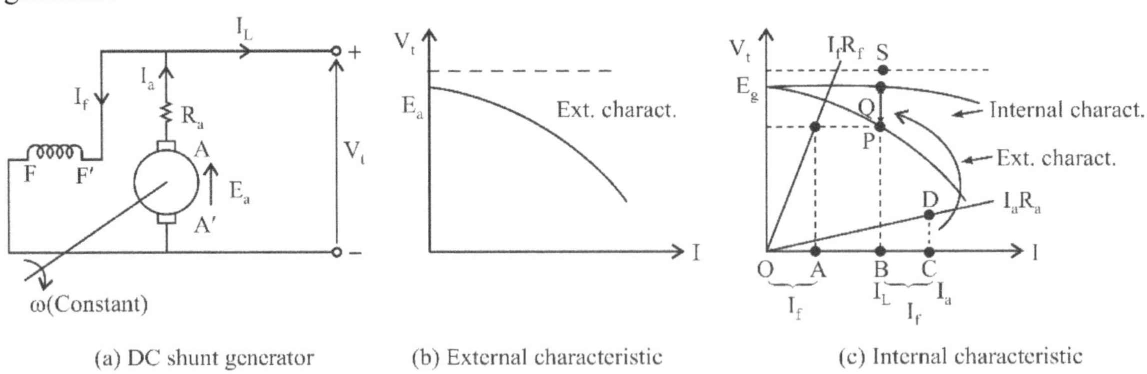

Figure 4.27 Internal and external characteristics of DC shunt generator

The external characteristic is plotted between V_t and I_L. When the load current is I_L (OB), the corresponding field current is OA (obtained by I_fR_f line). In order to obtain I_a which is $I_f + I_L$, OA is added to OB to give $I_a = OC$. At This value of I_a, the $I_a R_a$ drop (obtained by I_aR_a line) is obtained and the same is added to BP (terminal voltage) to get Ea equal to BQ. Here SQ is the drop in voltage due to armature reaction.

The above process is repeated for several points on the external characteristic to obtain corresponding points of internal characteristic. All such points are then added by a smooth curve to finally obtain the internal characteristic.

The operation of DC shunt generator becomes unstable after a certain value of I_L. As I_L increases the terminal voltage V_t drops down. With the decrease in V_t, the field current I_f becomes less. Reduction in I_f causes low flux and hence reduced emf generated. This leads to further decrease in V_t.

So, as $I_L \uparrow \rightarrow V_t \downarrow \rightarrow I_f \downarrow \rightarrow E_a \downarrow \rightarrow V_t \downarrow \rightarrow I_f \downarrow \rightarrow \phi \downarrow \rightarrow E_b \downarrow \rightarrow V_t \downarrow ......$

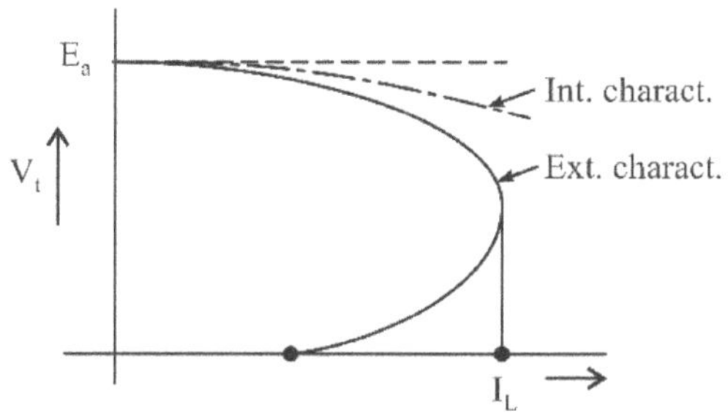

Figure 4.28 Collapse of terminal voltage

This process progresses uncontrollably, till it leads to zero voltage across the generator terminals with no induced emf available.

In other words, the emf de-builds itself and finally gets diminished.

4.14.3 Self Excited DC Series Generators

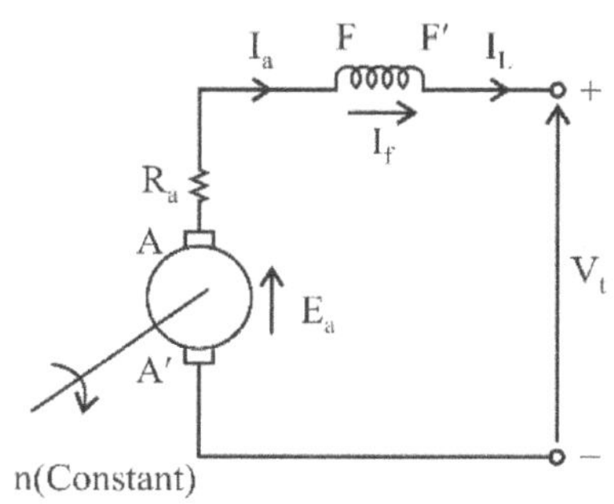

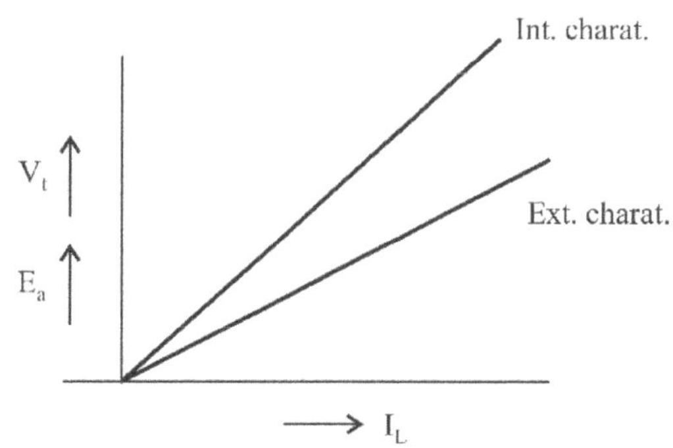

Figure 4.29 Internal and external characteristics of DC series generator

Here, $V_t = E_a - I_a (R_a + R_f)$

$I_a = I_f = I_L$

Increase in I_L results in increase of flux leading to increase in induced emf E_g. Hence, V_t increases with increase in I_L. After plotting external characteristic (V_t vs I_L) the $I_a(R_a + R_f)$ drop is correspondingly added to it and the internal characteristic is obtained.

4.14.4 DC Compound Generators

The field winding is split into two parts. One part is connected in series and the other in parallel with the armature winding. By doing so, intermediate characteristics can be obtained between series and shunt generator characteristics. The two parts of field winding can be connected in two different ways, both magnetically as well as electrically. Magnetically, one part of the field winding can be physically placed in a manner so as to either strengthen or weaken the field, produced by the second part of the field winding. Accordingly, the compound generator is called cumulatively compounded or differentially compounded generator respectively.

Similarly, there can be two ways of electrically connecting the two parts of field winding. In one connection the parallel combination of armature winding with one part of field winding is connected in series with the other part of the field winding. Generators having such connection are called Short shunt compound generators. In the other type of connection, the series combination of armature winding with one part of field winding is connected in parallel with the other part of the field winding. Such generators are called Long shunt compound generators.

Both the short compounded and long compounded generators can be either cumulatively compounded or differentially compounded. Fig.4.30 shows the four possibilities of connections of compound generators.

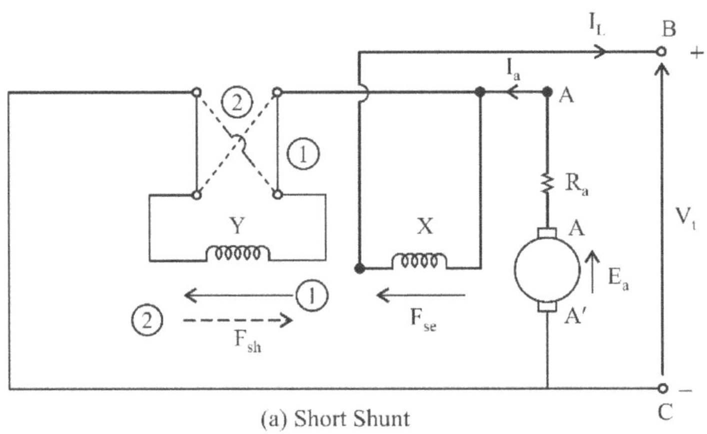

(a) Short Shunt

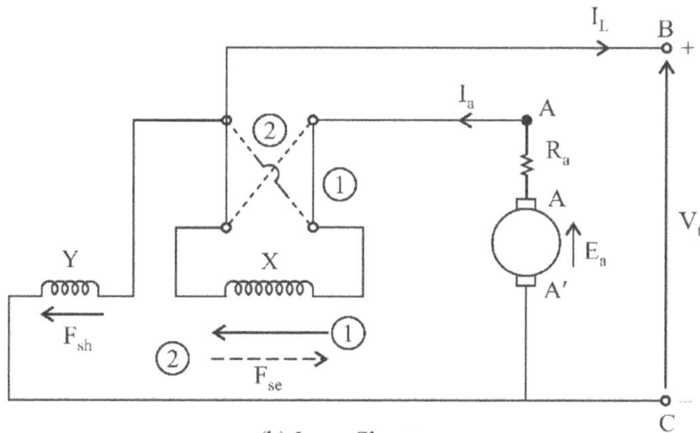

(b) Long Shunt

Figure 4.30 DC compound generators

The connections shown in Fig.4.30 by solid lines as (1) lead to cumulative compounding while those shown by discontinuous lines as (2) lead to differential compounding.

In cumulative compounding, the series and shunt components of field flux (F_{se} and F_{sh}) act in same direction, while in differential compounding, the series and shunt components act in opposite directions.

By changing the number of turns of series field winding the cumulatively compounded DC generators can act as over-compounded, flat-compounded or under-compounded generators. The terminal voltages that may be obtained from DC compound generators under no-load condition, are shown in Fig.4.31.

1. Over compounded
2. Flat-compounded
3. Under-compounded
4. Differentially compounded

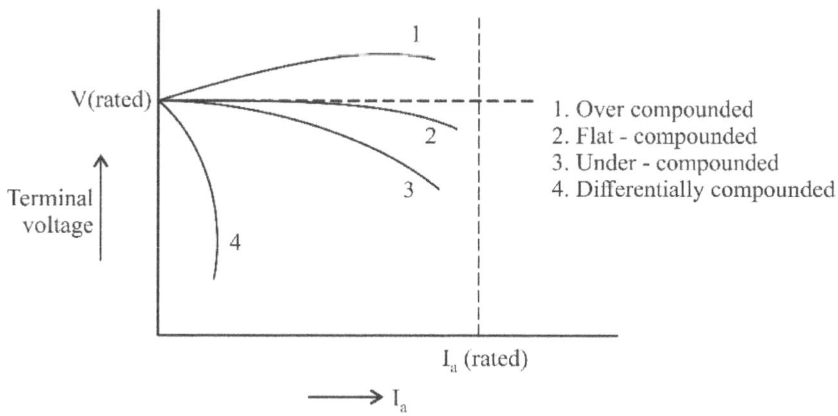

Figure 4.31 Characteristics of compound generator

Thus, a cumulatively or differentially compounded DC generator can be connected either as short shunt or long shunt to give the desired characteristic.

4.15 Losses and Efficiency of DC Generators

The power fed as input does not get fully converted into output power. Several losses take place in DC generators at various places. These losses are the copper loss, the iron loss and the mechanical loss.

4.15.1 Copper Loss (I^2R Loss)

The currents flowing through the windings cause the copper losses to take place. It is represented by multiplication of the square of the current and the resistance of the corresponding winding. The armature winding carries load current. Hence, this current square multiplied by armature resistance becomes the copper loss in armature winding. The copper loss in the shunt field winding is the multiplication of the terminal voltage multiplied by the current in field winding. The currents through series field winding, interpole winding, and compensating winding, decide the amount of copper loss caused by them. This is obtained by multiplying the series field current with resistances of corresponding windings.

Copper losses vary with the square of current.

4.15.2 Iron Loss

Iron losses occur when an ac current and the corresponding ac flux are present in a machine. The armature current in a DC generator is ac only which is rectified later by commutator-brush arrangement. The presence of this ac current in armature winding leads to development of hysteresis and eddy current losses. These losses are represented as follows.

Hysteresis loss = $k_h\ B_m^{1.6}\ f\ V_c$ watts and

Eddy current loss = $k_e\ B_m^2\ f^2\ t^2\ V_c$ watts

Where, k_h and k_e are the constants, B_m is the maximum flux density, f is the frequency, t is the thickness of core lamination and V_c is the volume of the core.

Iron losses are constant losses for constant speed and constant flux density of a machine. Iron losses occur in magnetic materials which experience pulsating alternating flux. As the current induced in armature of a DC machine is alternating, hence the armature core suffers from eddy current loss. The armature reaction flux is also pulsating in nature, and it interacts with poles, hence this phenomenon also leads to eddy current loss.

4.15.3 Mechanical Loss

Mechanical loss comprises of fictional and windage losses. The bearing, brushes and windage cause friction in movement of rotor. The brush and bearing losses are proportional to speed while the windage loss is proportional to the cube of speed of the rotor.

4.15.4 Efficiency

Like any other machine, the efficiency of a DC machine is defined as the ratio of its output power to input power.

$$\eta_G = \frac{P_{op}}{P_{ip}} = \frac{P_{op}}{P_{op} + \text{losses}} = \frac{P_{ip} - \text{losses}}{P_{ip}}$$

It can be derived that the maximum efficiency of the machine occurs when the variable losses are equal to constant losses.

4.16 Voltage Regulation of DC Generators

The voltage regulation of a DC generator can be termed as the rise in terminal voltage from full load to no load condition at rated speed. The external characteristic of a DC generator reveals its nature of voltage regulation as given in Fig.4.32.

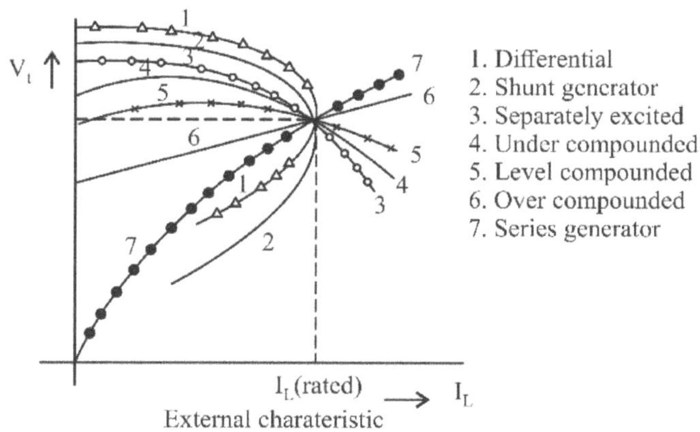

Figure 4.32 External characteristic of DC generators

The curves shown in Fig.4.32 show the external characteristics of various DC generators at constant (rated) speeds. These characteristics vary with variation in operating speeds.

The voltage regulations of DC shunt generators and cumulatively compounded generators are most susceptible to change in speeds. This phenomenon is explained for a shunt generator in Fig.4.33.

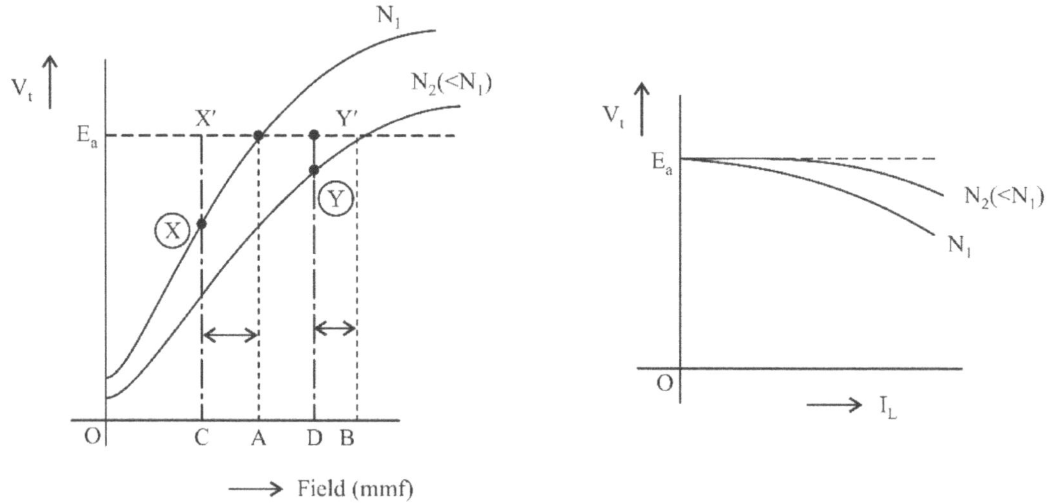

(a) OCC of shunt generator (b) External characteristic of shunt generator

Figure 4.33 Effect of speed on voltage regulation

In Fig. 4.33(a), the OCC are drawn for two speeds, N_1 and N_2 (where N_2 is less than N_1). Say, the point of operation is X at speed N_1. Here, the drop $X' - X$ is caused by armature reaction. The armature reaction is prominent in case of weaker magnetic fields, whereas the effect of armature reaction becomes less when the magnetic field is stronger in the generator. Now, say the speed is reduced from N_1 to N_2. The corresponding OCC will be as shown in Fig. 4.33(a). For same load current, the drop due to armature reaction will be less. Hence, the point of operation gets fixed at Y. It can be seen that the drop in terminal voltage is less at lower speed for the same load current $\left[(r-r')<(x-x')\right]$. The same is depicted in Fig. 4.33 (b). It is further of interest to see that the amount of mmf required to pull the terminal voltage to E_a is less, when the speed is lower [DB < CA].

4.17 Parallel Operation of DC Generators

Generators are required to operate in parallel when the load gets increased or when standby generator is required. The self-excited DC generators are connected in parallel whereas it is pointless to connect separately excited generators in parallel as they require separate sources for their excitation.

The essential conditions for generators to be connected in parallel are as follows.

1. They should have same terminal voltage.
2. They should be connected with correct polarity.
3. Their voltage regulation should be the same (as far as possible).

4.17.1 Parallel Operation of DC Shunt Generators

Say generator G_1 was supplying some load and now generator G_1 was supplying some load and now generator G_2 is to be added to the bus bar to share the load. Connection diagram shows the arrangement for the same.

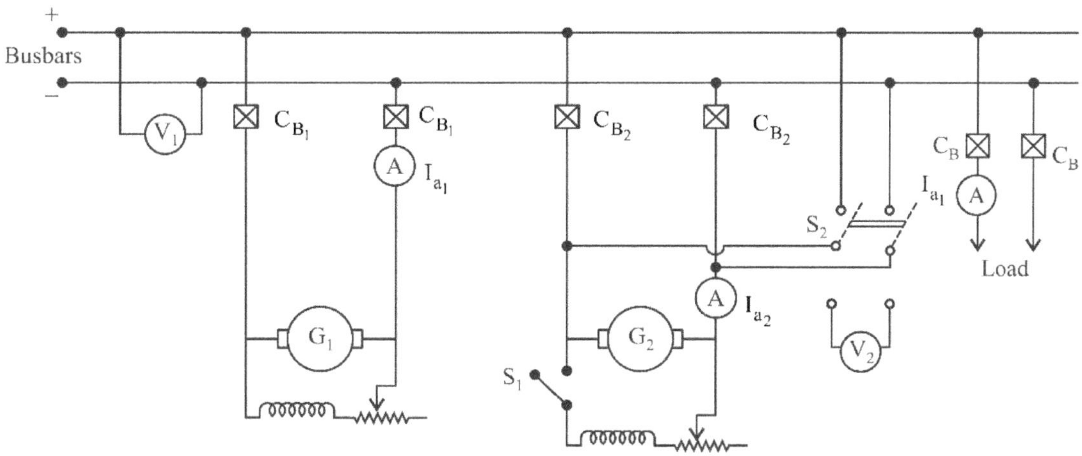

Figure 4.34 Parallel operation of DC shunt generators

In order to bring generator G_2 in parallel with G_1 (which is already supplying the load), the following steps are taken.

1. CB_1 are already closed and G_1 is feeding the load at voltage V_1.
2. CB_2, S_1 and S_2 are open.
3. Prime mover of G_2 is run at rated speed.
4. S_1 is closed and field of G_2 is excited. S_2 is switched to connect voltmeter V_2.
5. Excitation of G_2 is adjusted such that $V_2 = V_1$.
6. CB_2 is closed to bring G_2 into the circuit.
7. Initially no current flows through G_2.
8. Field of G_1 is slightly weakened. This will enable both the generators to feed the load.

4.17.1.1 Load Sharing

In order to understand the load sharing, the circuit shown in Fig. 4.35 is considered where field windings are ignored.

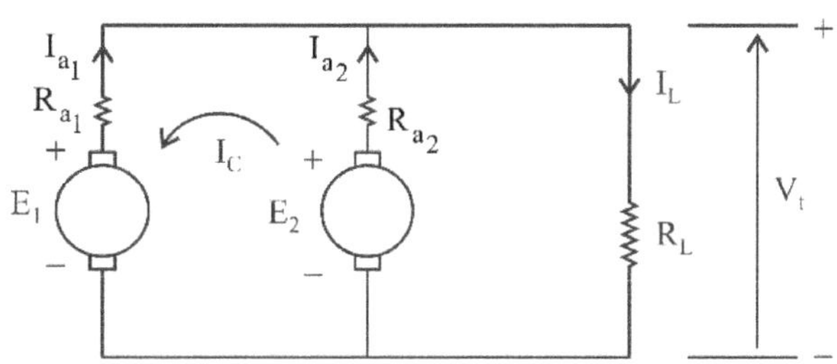

Figure 4.35 Load sharing in parallel operation

The circulating current I_2 is absent when the terminal voltages of G_1 and G_2 are equal making $I_{a_1} + I_{a_2}$ and $I_L = I_{a_1} + I_{a_2}$.

Equations governing external characteristics are as follows.

$$V_t = \left(E_1 - I_{a_1} R_{a_1}\right) = \left(E_2 - I_{a_2} R_{a_2}\right);\ I_L = I_{a_1} + I_{a_2}$$

$$\therefore \quad I_{a_1} = \left|\frac{E_a - E_1}{R_{a_1} + R_{a_2}}\right| + \frac{R_{a_2}}{R_{a_1} + R_{a_2}} I_L$$

$$I_{a_2} = \left|\frac{E_1 - E_2}{R_{a_1} + R_{a_2}}\right| + \frac{R_{a_1}}{R_{a_1} + R_{a_2}} I_L$$

[A] [B]

In the above equations of I_{a_1} and I_{a_2}, the components [A] represent the circulating current I_c which is independent of the load current I_L. The components [B] give the values of currents shared by the two generators feeding the load. It is clear that the circulating current I_c will be equal to zero if $E_1 = E_2$ and then load will be equally shared, if $R_{a_1} = R_{a_2}$.

4.17.1.2 Role of External Characteristics in Parallel Operation

For a given terminal voltage (V_t) and load current, the share of currents is governed by the external characteristics of the two generators running in parallel. Two different cases are depicted in Fig.4.36.

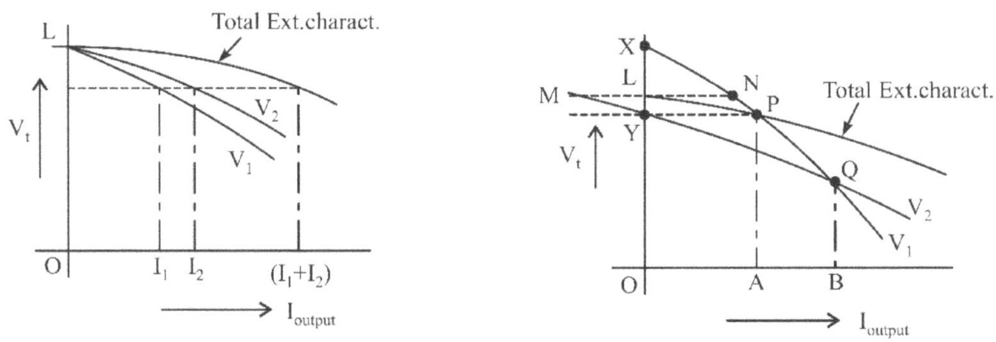

(a) When OC terminal voltages of G_1 and G_2 are the same (OL)

(b) When OC terminal voltages of G_1 and G2 are different (OX and OY)

Figure 4.36 External characteristics on Parallel Operation

Fig. 4.36(a) shows that the open circuit voltages of G_1 and G_2 are the same. Hence no circulating current flows when the generators run in parallel. In Fig. 4.36 (b), the two generators have different values of open circuit voltages (equal to OX and OY as shown). Consider the no load condition (at operating point L). G_1 generates current equal to LN, and the same amount of current LM is absorbed by G_2 (which therefore starts acting as a motor). This current keep circulating between the two machines under no load condition. As the load current rises above zero, say the operating point shifts to P. Here the load current OA is shared only by G_1 and current through G_2 is zero. Now, with further increase in load current, the operating point shifts to a position Q such the currents shared by both the generators are equal (OB) and hence no circulating current would be there. So, it is seen that if the no load terminal voltages of G_1 and G_2 are unequal, then one of the generators acts as a motor. With increasing load current up to OA, only G_1 supplies to the load. A further increase in load current from A to B, share of G_1 remains more than that of G_2. When the load current rises beyond OB, the share of G_2 becomes more than that by G_1.

It is clear from above discussion that the drooping characteristic of shunt generators helps in maintaining their stable parallel operation.

4.17.2 Parallel Operation of DC Series Generators

Due to rising characteristics of DC generators it is difficult to run them in parallel to share a load. Hence, they are usually not operated in parallel. When the two DC series generators are put in parallel to supply a load, an increase in current in one of the generators will lead to increase in flux and hence the emf of that machine. This will result in further increase in current. This cumulative process will make the other machine in parallel to absorb current. Thus, the first machine will start supplying to a parallel combination of the load resistance and a very small resistance of the second machine with an additive polarity. This will result in progressively rising flow of current between the two generators leading to short circuit condition.

However, there can be two ways to connect the two DC series generators in parallel.

(a) with cross field connection

(b) with equalizer connection

The two connections are shown in Fig.4.37

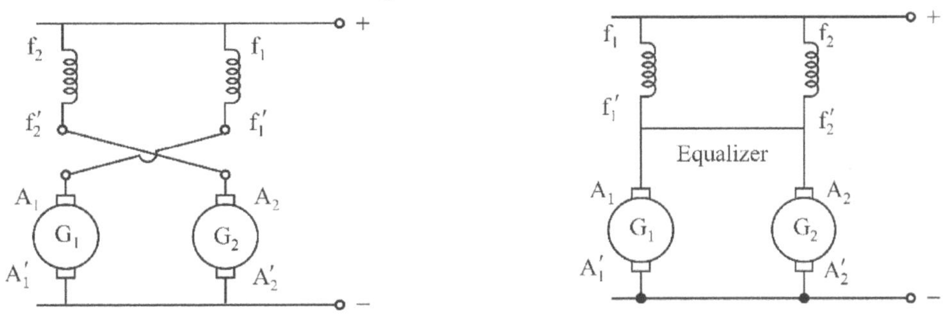

(a) Cross field connection

(b) Equalizer connection

Figure 4.37 Parallel operation of DC series generators

In case (a), any increase in field current of one of the machines will increase the emf of the other machine and vice versa. This will restrict the progressive rise in any of the quantities. Similarly, in case (b), the equalizer connection will check the field currents in a manner so that the emfs of the two machines respond to rise in load current in a stable manner.

4.18 Applications of DC Generators

Applications of DC generators are shrinking due to their cost, maintenance and other technical reasons. DC generator applications are mainly based on their characteristics.

(a) DC separately excited generators: These generators are expensive as they require separate DC source for excitation. They can provide wide range of terminal voltage at output. Hence, they are generally used in laboratories for testing purpose. Sometimes these generators are used to supply such DC motors which are required to operate the controlled speed loads.

(b) DC series generators: DC series generators have rising terminal voltage characteristics with load current. However, they can provide constant supply to load. They find application in DC locomotives where regenerative breaking is applied. They can act as booster where voltage drop occurs in long feeders.

(c) DC shunt generators: DC shunt generators have nearly constant terminal voltages having slight droop. They are generally used for battery charging, portable generators, lighting, providing excitation to alternators etc.

(d) DC Compound Generators: DC Compound generators are the most widely used among DC generators as their external characteristics can be made flexible by adjusting the series field winding to match the output voltage requirement of loads. The compounding levels in cumulatively compounded generators may be made over, level or under compounded. Cumulatively compounded DC generators suffer heavy demagnetization armature reaction hence their characteristics are suitable for arc welding.

REVIEW QUESTIONS

1. Describe the following parts of a DC Machine. Also tell their roles.
   (i) Poles (ii) Slots,
   (iii) Commutators, (iv) Brushes,
   (v) Armature winding
2. Which are the two types of armature windings in DC machine? Compare them using the following.
   (i) Pole pitch, (ii) Back pitch,
   (iii) Front pitch, (iv) Commutator pitch,
   (v) Diagrams showing their layout
3. Develop from first principle, the EMF and Torque equations for a DC machine.
4. Explain with the help of OCC and R_f line, the process of development of emf in a DC shunt machine. Also explain the term 'Critical Speed' and its significance.
5. In what manner does the armature current, interact with the main field flux? With the help of a suitable diagram explain as to why it is undesirable. What is done to overcome it?

PROBLEMS

1. Design a suitable winding for a 4-pole armature with 2540 conductors, 127 commutator segments and 32 slots.
2. Find the back pitch y_b, front pitch y_f, resultant pitch y_r and average pitch y_a for a 34-coil side wave wound d.c. machine having 6 poles. Also give the scheme of connecting the conductors.

3. A 6-pole machine has an armature with 90 slots having 8 conductors per slot and runs at 1000 rpm. The flux/pole is 0.05 Wb. Determine induced emf if the winding is:

   (i) Lap connected (ii) Wave connected

4. A 6-pole DC machine has 60 slots having 6 conductors/slot and each conductor carries a current of 50 A. The machine core diameter is 0.4 m, length is 0.3 m and the average value of the flux density is 0.3T. The machine is driven at 1500 rpm. Determine emf, armature current, power developed, electro-magnetic torque developed and the flux in field yoke, if armature conductors are:

   (a) Wave connected (b) Lap connected

5. A 250 volt compound generator is operated with a long shunt and then with a short shunt connection. Its series field and armature resistances and shunt field resistance are 0.1 ohm, 0.4 ohm and 150 ohms respectively. Ignore armature reaction. Take 1 V per brush as the contact drop. Find the values of induced emfs for both the connections.

CHAPTER 5
DC MOTORS

5.1 Working Principle of DC Motors

A DC motor is a machine that converts DC electrical energy into mechanical energy.

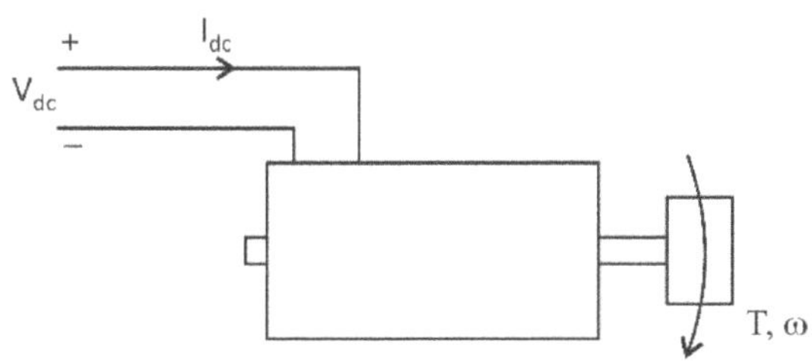

Figure 5.1 DC Motor

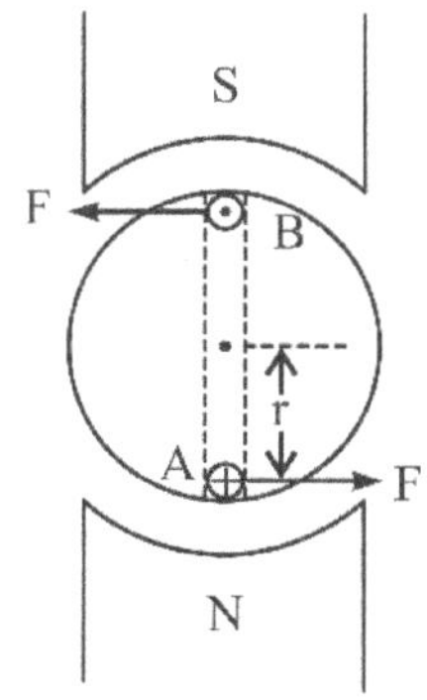

Figure 5.2 Torque development

Consider a DC motor running at speed (ω) with reference and torque T Nm. The output power is Tω watt. The input dc voltage is V_{dc} and input current is I_{dc}. Input power is V_{dc} I_{dc} watt. The input dc voltage is V_{dc} and input current is I_{dc}. Input power is V_{dc} I_{dc} watt. Fig.5.1 gives an idea that the dc input energy is used to run the motor and mechanical energy can be utilized in driving the load. The DC motors have many good qualities like high starting torque, wide range of speed control and speed reversals. In construction there is no difference between a DC motor and DC generator. The same machine can operate as a generator or a motor. To understand the principle of working of the DC motor, consider a coil lying at the center of N and S poles. Let the current flowing be downwards in conductor A and upwards in conductor B. The direction of the field is upward. Applying Flemings left hand rule for motor action, it is observed that a force is developed to move the conductor from right side and similarly a force is developed to move the conductor B on left side. Thus, these two forces shall try to rotate the motor in anti-clockwise direction about the center of the coil. In a practical DC motor these are large number of coils and are so placed that the torque developed in all of them shall tend to rotate the armature in same direction.

5.1.1 Production of Torque

Consider a current carrying coil AB lying in magnetic field as shown in Fig. 5.2. By Fleming's Left Hand Rule the force developed by coil AB shall tend to turn it in anti-clockwise direction. Force Experienced by conductor A.

$$F = BI_cL \text{ (N)} \qquad .....(5.1)$$

where

B = Magnetic flux density (T)

$I_c = \frac{I_a}{a}$ conductor current (A)

L = Length of conductor

I_a = Armature current

a = No. of parallel paths (wave = 2, Lap = 2p)

$2p$ = No. of poles

Conductor current in A is downwards and upwards in conductor B. The direction of the field is upwards. Applying Flemings Left Hand Rule for motor action, it is observed that a force is developed to move the conductor A from right side and similarly a force is developed to move the conductor B on left side. Thus, these two forces shall try to rotate the motor in anti-clockwise direction about the center of the coil. In a practical DC motor there are large number of coils and are so placed that the torque developed in all of them shall tend to rotate the armature in same direction.

5.2 Production of Torque

Consider a current carrying coil AB lying in magnetic field as shown in Fig.2. By Flemings left hand rule the force developed by coil A-B shall tend to turn it in anti-clockwise direction. Force experienced by conductor A.

$$F = B\, I_c\, L \text{ (N)} \qquad .....(5.2)$$

where

B = Magnetic flux density (T)

$I_c = \frac{I_a}{a}$ = conductor current (A)

I_a = Armature current (A)

a = No. of parallel paths (wave = 2, Lap = 2p)

L = Length of conductor (ml)

$2p$ = No. of poles

Torque developed by one conductor at A

$$B = F \times r$$

$$= B\, I_c\, L \times r \text{ (Nm)} \qquad ....(5.3)$$

where r = radius (m)

If Z is total number of conductors in the armature, the total torque developed is

T = No. of conductors × Torque developed by one conductor

$$= Z \times B_c L \times r \quad (Nm)$$

$$= Z \times BL\, r \times I_c \quad (Nm) \qquad .....(5.4)$$

Let ϕ be flux/pole (wb),

$$\phi = \text{Area of one pole surface} \times B$$

$$= \frac{2\pi rL}{2p} \times B$$

or

$$BL\, r = \frac{p\phi}{\pi} \qquad .....(5.5)$$

Putting the value of BL r in equation 5.4

$$T = Z \times \frac{p\phi}{\pi} \times I_c \qquad .....(5.6)$$

Putting the value of I_c from equation 2

$$T = Z.\, \frac{p\phi}{\pi} \,.\, \frac{I_a}{a}$$

$$\frac{pZ}{\pi A} \,.\, \phi \,.\, a$$

$$= K_t\, \phi\, I_a \qquad ....(5.7)$$

where

$$K_t = \frac{pZ}{\pi a} \qquad .....(5.8)$$

K_t is constant for a given machine

Example 1: The armature of a 6-pole lap wound DC shunt motor takes 40 A. The flux/pole is 80 mW_b and number of turns is 600. Determine the torque developed and also find if the motor is wave wound.

Solution: $p = 6$, $I_a = 40A$, $\phi = 80 \times 10^{-3}$ wb

No. of turns = 600, $Z = 2 \times 600 = 1200$

Lap wound

$$T = K_t\, \phi\, Ia$$

$$K_t = \frac{pZ}{\pi a} = \frac{3 \times 1200}{\pi \times 6} = 191$$

$$T = 191 \times 80 \times 10^{-3} \times 40 = 611 \text{ Nm}$$

Wave wound

$$K_t = \frac{pZ}{\pi a} = \frac{3 \times 1200}{\pi \times 2} = 573$$

$T = 573 \times 80 \times 10^{-3} \times 40 = 1833$ Nm

For Lap winding A = 6 and wave winding a = 2

5.3 Counter or Back EMF

When the armature of a DC motor rotates, the armature conductors cut the flux. As per Faraday laws of electro-magnetic induction there is induced emf in the, armature conductors which opposes the applied voltage.

For the armature circuit of DC motor, the EMF equation is

$$Vt = E_b + I_a R_a \quad \text{......(5.9)}$$

Or

$$E_b = V_t - I_a R_a \quad \text{.....(5.10)}$$

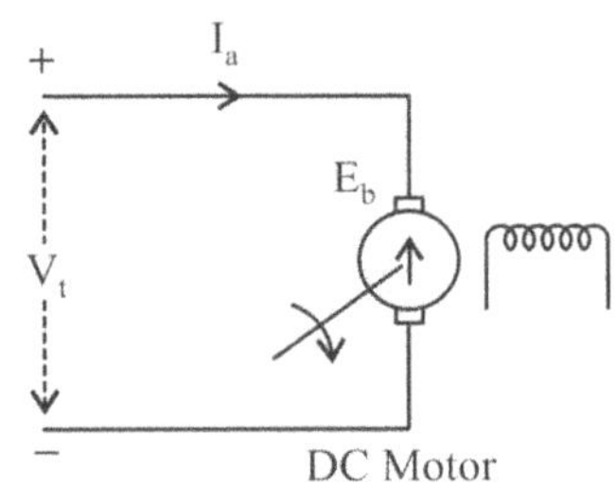

Figure 5.3

where

V_t = Applied dc voltage (v)

E_b = Back EMF (V)

I_a = Armature current (A)

I_a = Armature resistance (Ω)

$I_a R_a$ is armature circuit resistance drop. It is similar to internal resistance drop in a Ballery circuit.

From equation 5.7

$$I_a = \frac{V_t - E_b}{R_a} \quad (A) \quad \text{.....(5.11)}$$

The Back EMF is nothing but an induced emf in the armature of the DC motor due to its rotation and is exactly the same as we get in a DC generator.

The equation for back emf may be given by

$$E_b = K_n \phi N \text{ (V)} \quad \text{.....(5.12)}$$

where

$$K_n = \frac{pZ}{60\ a}$$

P = number of poles

a = number of parallel paths

ϕ = flux/pole (wb)

N = speed (rpm)

5.4 Types of DC Motors

The DC motors may be classified as under:

1. Separately excited motor

2. Self-excited motor. Further classified as:
   (a) Shunt motor
   (b) Series motor
   (c) Compound motor
      (i) Short shunt cumulative
      (ii) Short shunt differential
      (iii) Long shunt cumulative
      (iv) Long shunt differential

Their diagrammatic representation and characteristics discussed here under.

5.4.1 Separately Excited DC Motors

The electrical representation of a separately excited DC motor is given in Fig.5.4. In this case armature and field are supplied from independent sources. The basic equations are:

$$E_b = V_t - I_a R_a \text{ (V)} \quad .....(5.13)$$

$$I_f = \frac{V_f}{R_f} \text{ (A)} \quad .....(5.14)$$

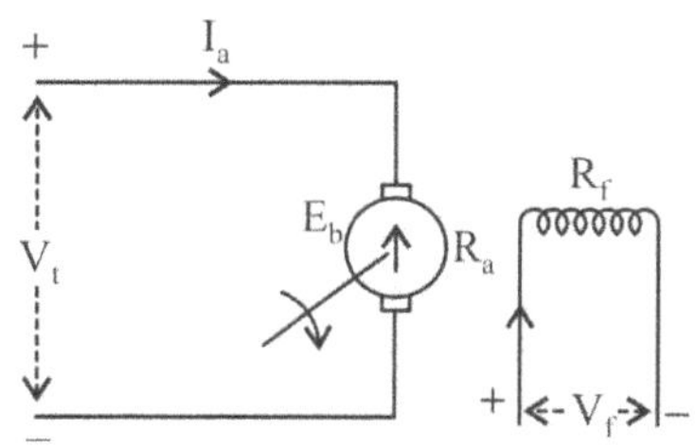

Figure 5.4

where V_f = dc field voltage (V)

R_f = Field winding resistance (Ω)

I_f = Field current

5.4.2 Self Excited DC Shunt Motors

The electrical representation of a DC shunt motor is shown in Fig.5.4. The shunt winding is connected in parallel with the armature.

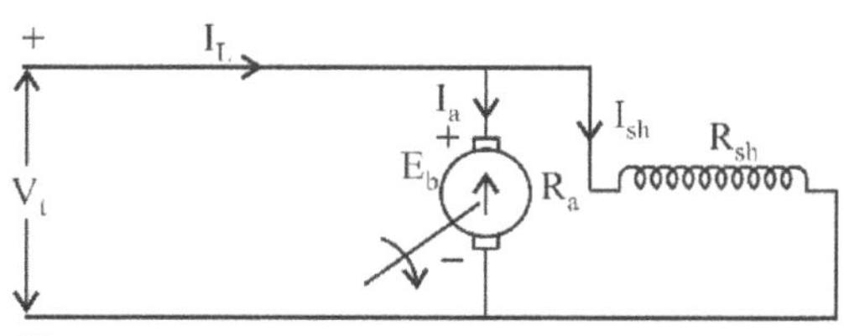

Figure 5.5 DC Shunt Motor

Here the input current I_L divides into two parts. Current I_a through the armature and I_{sh} through the shunt field winding. The equations are:

$$E_b = V_t - I_a R_a \quad .....(5.15)$$

$$I_L = I_a + I_{sh} \quad .....(5.16)$$

$$I_{sh} = \frac{V_t}{R_{sh}} \quad .....(5.17)$$

where I_{sh} is shunt field current and R_{sh} shunt winding resistance.

5.4.3 Self Excited DC Series Motors

The electrical representation of DC series motor is given in Fig. 5.6. In this case the field is in series with the armature.

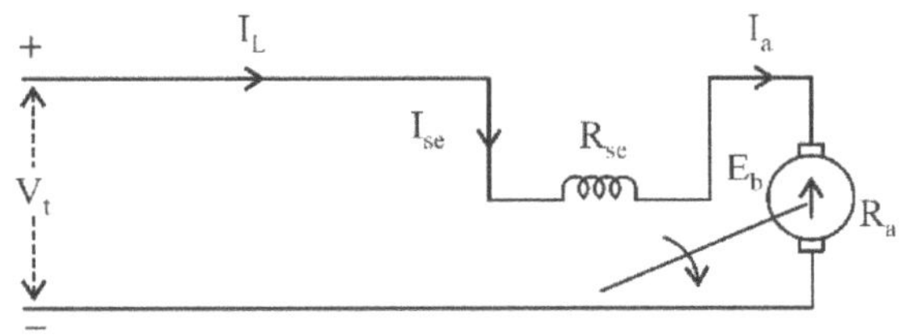

Figure 5.6 DC series motor

The relevant equations are

$$E_b = V_t - I_a (R_a + R_{se}) \quad .....(5.18)$$

$$I_L = I_a = I_{se} \quad .....(5.19)$$

where I_{se} is series field current and R_{se} is series field resistance.

5.4.4 DC Compound Motors - Short Shunt Cumulative

The electrical representation of a short shunt cumulatively compounded motor is shown in Fig. 5.7.

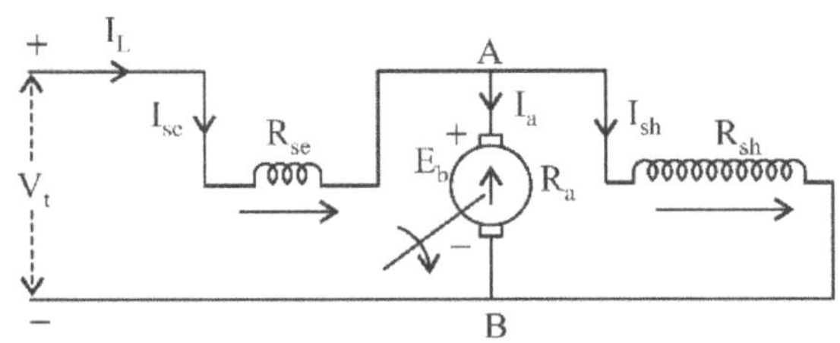

Figure 5.7 DC short shunt cumulatively compounded motor

In this the series field helps the shunt field, which is taken as the main field. Its effect is that as the load increases, the series field flux increases, which adds to the shunt field flux. The equations are:

$$E_b = V_t - I_a R_a - I_{se} R_{se} \quad .....(5.20)$$

$$I_{sh} = \frac{V_{AB}}{R_{sh}} \quad .....(5.21)$$

$$V_{AB} = V_t - I_{se} R_{se} \quad .....(5.22)$$

$$I_L = I_{se} = I_a + I_{sh} \quad .....(5.23)$$

5.4.5 DC Compound Motors - Short Shunt Differential

The electrical representation of a short shunt differential compound motor is shown in Fig.5.8.

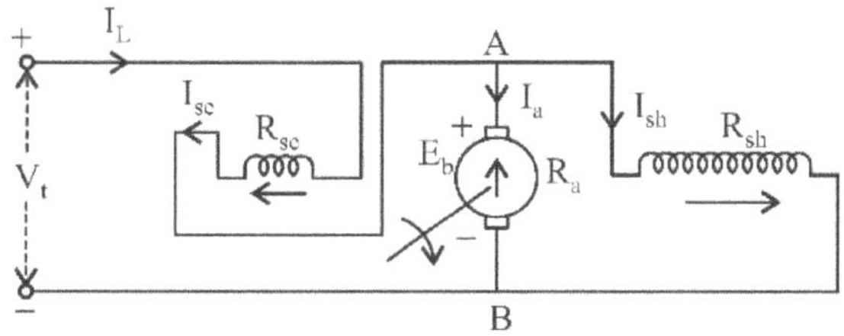

Figure 5.8 DC short shunt differentially compounded motor

In this case the field winding is so connected that its field opposes the main shunt field. Its effect is that as the load increases, the series field flux increases in opposite direction of the main shunt field flux. The net result is that the resultant flux decreases as the load increases.

The equations are:

$$E_b = V_t - I_a R_a - I_{se} R_{se} \quad .....(5.24)$$

$$I_{sh} = \frac{V_{AB}}{R_{sh}} \quad .....(5.25)$$

$$V_{AB} = V_t - I_{se} R_{se} \quad .....(5.26)$$

$$I = I_{se} = I_a + I_{sh} \quad .....(5.27)$$

5.4.6 DC Compound Motors-Long Shunt Cumulative

The electrical representation of a long shunt cumulative and differential compound motors are shown in Fig.5.9.

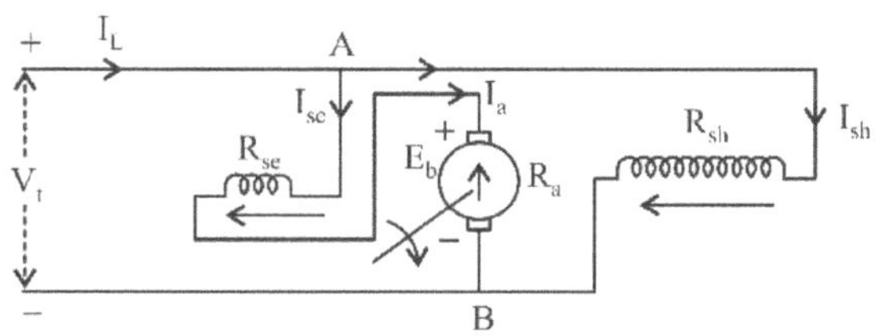

(a) Cumulative

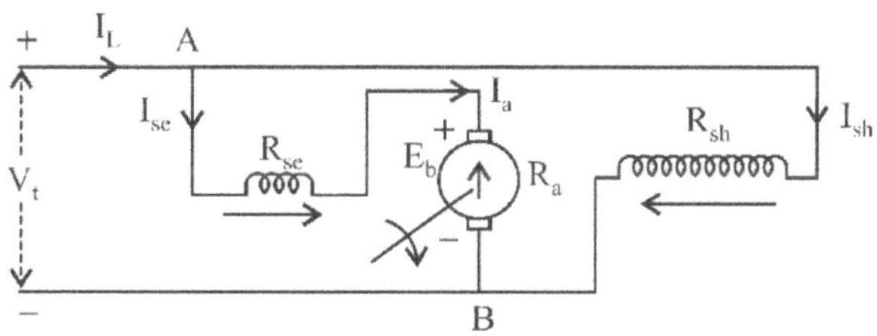

(b) Differential

Figure 5.9 DC long shunt compounded motors

Here the shunt field R_{sh} is connected in parallel with series field and the armature at points A and B. The equations are:

$$E_b = V_t - I_{se} R_{se} - I_a R_a \quad .....(5.28)$$

$$I_{sh} = \frac{V_{AB}}{R_{sh}} \quad .....(5.29)$$

$$V_{AB} = V_t \quad ....(5.30)$$

$$I_a = I_{se} = I_L - I_{sh} \quad .....(5.31)$$

In long shunt cumulative compound motors the two fields are additive, while in long shunt differential compound motors, the series field opposes the main shunt field.

5.5 Operating Characteristics of DC Motors

The following are important operating characteristics of DC motors:

1. Speed – Armature current characteristic ($N - I_a$)
2. Torque – Armature current characteristic ($T - I_a$)
3. Speed – Torque ($N - T$) characteristic

5.5.1 Characteristics of Separately Excited DC Motors

From equation 5.12,

$$E_b = K_n \phi N \quad .....(5.32)$$

$$N = \frac{E_b}{K_n \phi} \quad .....(5.33)$$

From equation 5.13, for separately excited motor

$$E_b = V_t - I_a R_a$$

Putting the value of E_b in equation 5.31

$$N = \frac{V_t - I_a R_a}{K_n \phi} \quad .....(5.34)$$

when $I_a = 0$

$$N = \frac{V_t}{K_n \phi}$$

For a separately excited DC motor, ϕ is constant; K_n is already a constant. So, for a given value of V_t, N is maximum. As I_a increases drop, the I_aR_a drop increases, decreasing the numerator of equation 5.34. Thus the speed also decreases as shown by ($N - I_a$) curve in Fig.5.10.

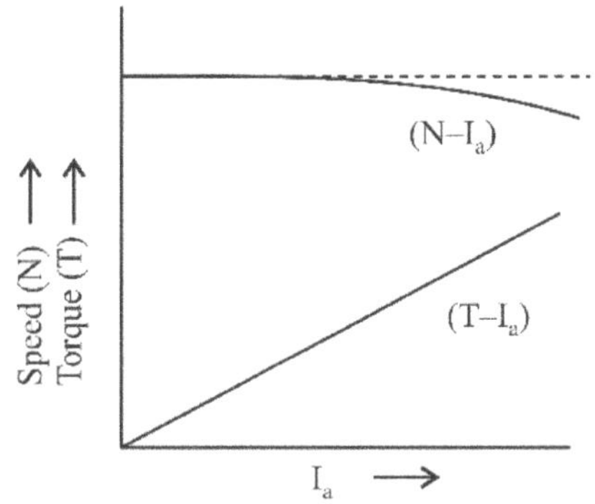

Figure 5.10 Speed torque characteristics against I_a

From equation 5.7, the torque is

$$T = K_t \phi I_a$$

K_t and ϕ are constants. When $I_a = 0$, $T = 0$. As I_a increases on loading, T also increases linearly as shown by $(T - I_a)$ curve in Fig.5.10.

Torque T v/s N characteristic could be derived from the two characteristics $(N - I_a)$ and $(T - I_a)$ of Fig.5.10. When $I_a = 0$, N is maximum and T is zero. As I_a increases there is a small drop is N. The Torque T versus N characteristic is shown in Fig.5.11.

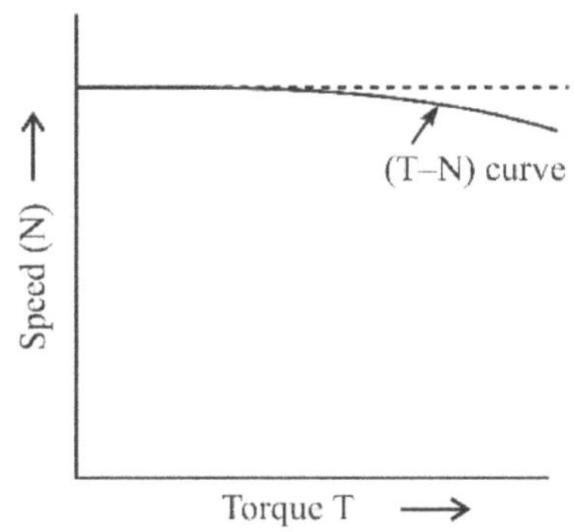

Figure 5.11 Speed – Torque characteristic

Example 5.2: A 200 V DC series motor has armature resistance of 0.06 Ω and series field resistance of 0.04 Ω. If the motor intake is 20 kW, determine back emf of the motor and power developed in the armature.

Solution: Motor input current I is

$$I = \frac{KW \times 10^3}{voltage}$$

$$= \frac{20 \times 10^3}{200} = 100\ A$$

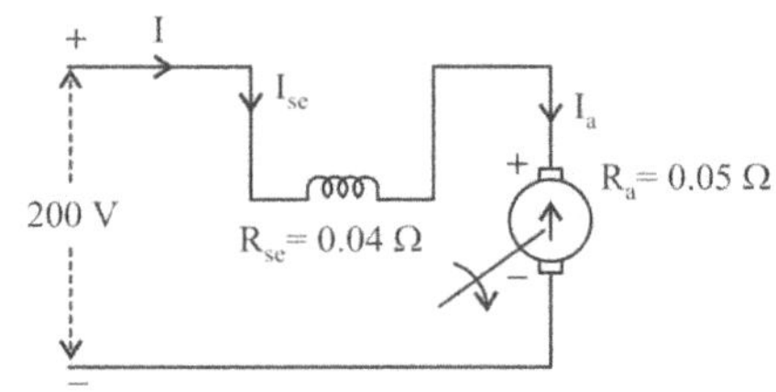

Figure A

Here $I = I_{se} = I_a = 100\ A$

Back emf of the motor

$E_b = V_t - I_a\ (R_a + R_{se})$

$= 200 - 100\ (0.06 + 0.04)$

$= 190\ V$

Power developed in the armature

$= E_b\ I_a$

$= 190 \times 100\ W$

$= 19\ kW$

Example 5.3: A 250 V, DC shunt machine has armature and shunt field resistances of 0.1 Ω and 125 Ω respectively. As a generator it delivers 20 kW output power at 250 V when running at 1000 rpm. Determine:

(a) The speed of the same machine when running as a shunt motor and taking an input of 20 kW at 250 V.

(b) The internal power developed when working as a:

(i) generator delivering 20 kW

(ii) motor taking 20 kW

Solution: In Generator as shown in Fig. B, the output current

$$I_L = \frac{20 \times 10^3}{250} = 80\ A$$

$$I_{sh} = \frac{250}{125} = 2\ A$$

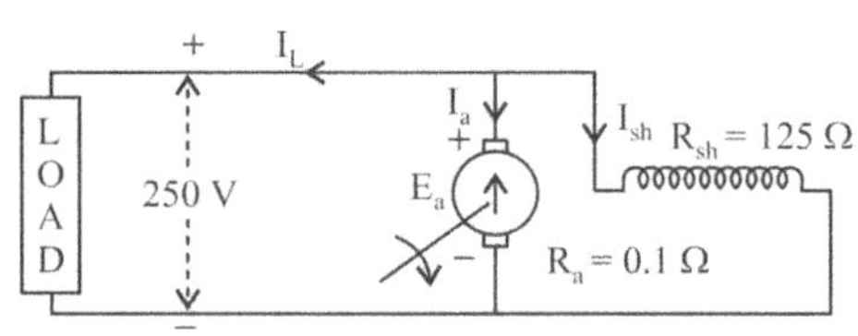

Figure B As generator

Armature current

$I_a = I_L + I_{sh}$

$= 80 + 2 = 82\ A$

Induced emf

$E_a = V_t + I_a\ R_a$

$= 250 + 82 \times 0.1$

$= 258.2\ V$

In Motor as shown in Fig. C, the input current

$$I_L = \frac{20 \times 10^3}{250} = 80 \text{ A}$$

$$I_{sh} = \frac{250}{125} = 2\text{A}$$

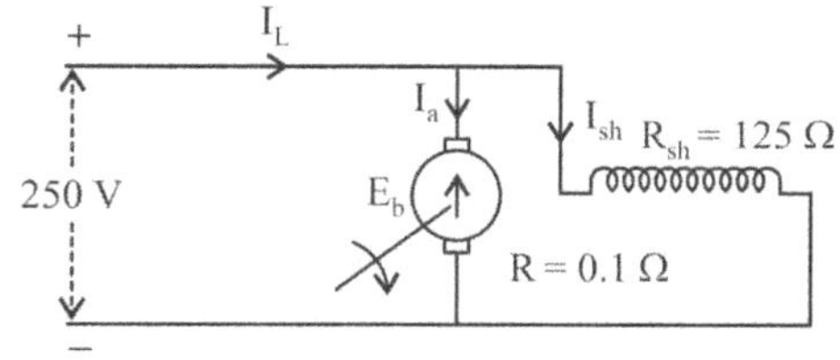

Figure C As motor

Armature current

$I_a = I_L - I_{sh} = 80 - 2 = 78$ A

Back emf

$E_b = V_t - I_a R_a$

$= 250 - 78 \times 0.1$

$= 242.2$ V

$$\text{Speed } N \propto \frac{E_b}{\phi}$$

Since I_{sh} is constant in both cases, ϕ is constant, so

$$N \propto E_b$$

Hence the speed of the N_{motor} is given by the following relationship

$$\frac{N_{motor}}{N} = \frac{E_b}{E_a}$$

Or

$$N_{motor} = \frac{E_b}{E_a} . N$$

$$= \frac{242.2}{258.2} \times 1000$$

$$= 938 \text{ rpm}$$

(b) (i) Internal power developed as generator

$= E_a \times I_a$

$= 258.2 \times 82 \times 10^{-3}$ kW

$= 21.17$ kW

`

(ii) Internal power developed as motor

$= E_b \times I_a$

$= 242.2 \times 78 \times 10^{-3}$ kW

$= 18.9$ kw

Example 5.4: A 4-pole 230 V, lap wound shunt motor takes a current of 52 A from the supply mains. The number of armature conductor is 600 and shunt field resistance is 115 Ω. The air gap diameter is 0.3 m, effective length of the pole is 0.2 m and average flux density in the airgap is 0.4 T. Determine the torque developed in the motor.

Solution:

Given: 2p = 4, V = 250 V, I_L = 52 A, Z = 600, R_{sh} = 115 Ω, air gap dia D = 0.3 m, effective length of pole l_{pole} = 0.2 m, B = 0.4 T.

Shunt field current

$$I_{sh} = \frac{230}{115} = 2\text{ A}$$

Armature current

$$I_a = I_L - I_{sh}$$

$$= 52 - 2 = 50\text{ A}$$

$$\text{Pole area} = \frac{\pi D}{2p}\, l_{pole}$$

$$= \frac{\pi \times 0.3}{4} = 0.2 = 0.0471/\text{m}^2$$

Flux/pole ϕ = Pole area × Flux density

$$= 0.0471 \times 0.4 = 0.0188\text{ wb}$$

$$K_t = \frac{Pz}{60a} = \frac{2 \times 600}{\pi \times 4} = 95.54$$

Torque developed

$$T = K\,\phi\,I_a$$

$$= 95.54 \times 0.0188 \times 50$$

$$= 90\text{ Nm}$$

5.5.2 Characteristics of Shunt Motors

Equations 5.12, 5.15, 5.16 and 5.17 give

$$E_b = K_n\,\phi\,N \quad .....(5.35)$$

$$E_b = V_t - I_a R_a \quad .....(5.36)$$

$$I = I_a + I_{sh} \quad .....(5.37)$$

$$I_{sh} = \frac{V_t}{R_{sh}} \quad .....(5.38)$$

From equations 5.35 and 5.36,

$$N = \frac{E_b}{K_n \phi}$$

$$= \frac{V_t - I_a R_a}{K_n \phi}$$

For shunt motor, ϕ is constant and all conditions of operation are similar to what exist for a separately excited DC motor. Hence the speed – armature current ($N - I_a$), Torque – armature current ($T - I_a$) and speed – torque ($N - T$) characteristics will be similar as shown in Fig.5.10 and 5.11.

Torque developed in the armature is $T = K_t \phi I_a$. The available torque is less than this torque after subtraction of the losses. The available torque is termed shaft torque or brake torque which is less than the armature torque (represented by dotted straight line) in Fig.5.12 (a).

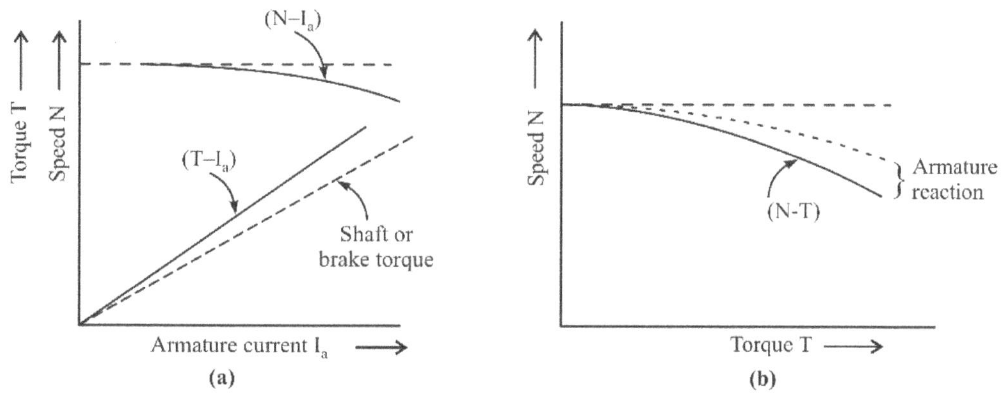

Figure 5.12 Speed torque characteristic

The speed-Torque ($N - T$) characteristic will also be similar to that of the separately excited motor as shown in Fig.5.12(b).

In shunt motors, the effect of armature reaction adds to effect of armature drop with increase in armature current. The drop in speed gets increased with increase in torque (and armature current).

5.5.3 Characteristics of DC Series Motors

Equations 5.12, 5.18 and 5.19 are,

$$E_b = K_n \phi N \quad .....(5.39)$$

$$E_b = V_t - I_a (R_a + R_{se}) \quad .....(5.40)$$

$$I_L = I_a = I_{se} \quad .....(5.41)$$

In equation 5.39 $I_a (R_a + R_{se})$ drop is small as compared to V_t, so,

$E_b = V_t$ (constant)

$\phi \propto I_{se}$

or $\phi \propto I_a$

So equation 5.37 may be written as

$$E_b = K_n I_a N$$

or $$I_a N = \frac{K_n}{E_b} \text{ (constant)} \qquad \text{.....(5.42)}$$

This equation is similar to xy = constant, which represents an hyperbolic curve. Hence the curve between speed N and armature current I_a shall be hyperbolic in nature as shown in Fig.5.13(a). At higher values of I_a, the field gets saturated, and hence the speed drops faster as shown by the dotted line in Fig.5.13(a).

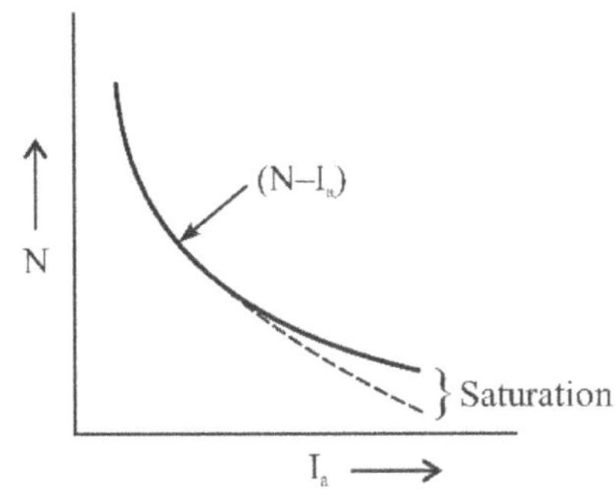

Figure 5.13(a) (N – I_a) Characteristic of Series motor

The torque equation is

$$T = K_t \phi I_a \qquad \text{.....(5.43)}$$

Till saturation of the magnetic circuit $\phi \alpha I_a$, so equation 5.41 give

$$T = K I_a^2 \qquad \text{.....(5.44)}$$

Equation 5.44 indicates that the armature torque developed is proportional to the square of the armature. Hence series motors have a supreme characteristic of developing very high torque which is useful for traction and locomotive etc.

When saturation is reached, ϕ becomes almost constant. Hence equation 5.41 gives

$$T \propto I_a \qquad \text{.....(5.45)}$$

Now the torque is proportional to armature current (T – I_a), the curve becomes linear. At start it is hyperbolic as given by equation 5.44. The nature of the armature torque T verses armature current I_a shall be as shown in Fig. 5.13(b).

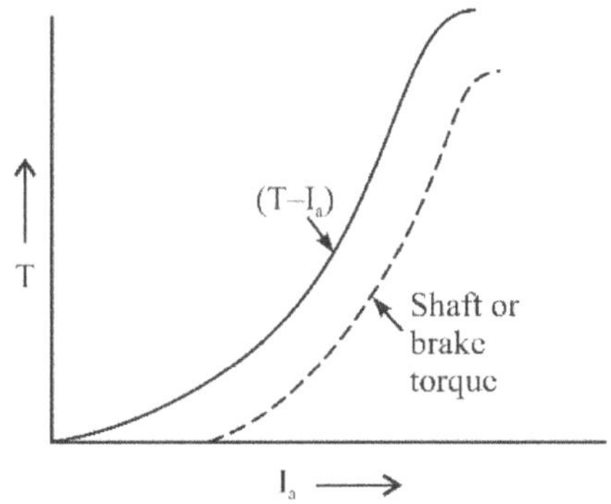

Figure 5.13(b) ($T - I_a$) characteristics of series motor

The speed torque characteristic of DC series motor is shown in Fig. 5.14.

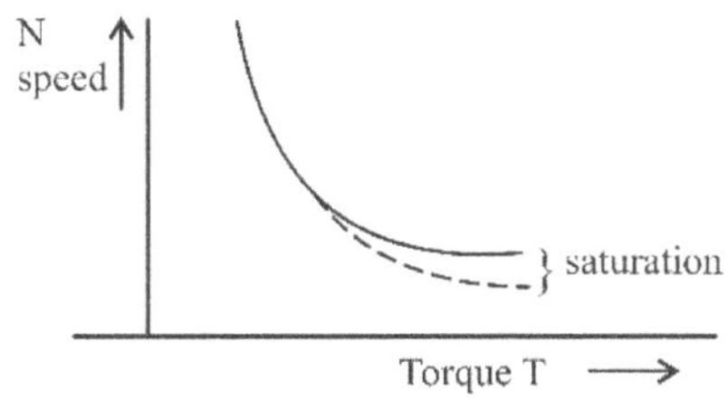

Figure 5.14 Speed-Torque Characteristic of Series motor

If the value of I_a is substituted in equation 5.42 from equation 5.43, we get

$$\left(\frac{T}{K_t\phi}\right)N = \frac{K_n}{E_b}$$

or $N\,T \propto \phi$ (as E_b is constant)

The motor usually runs near saturation, hence ϕ is also nearly constant. Thus, the N – T characteristic is more or less hyperbolic in nature as shown.

5.5.4 Characteristics of DC Compound Motors

Equations 5.7 and 5.12 give

$$T = K_t\,\phi\,I_a \quad .....(5.46)$$

$$E_b = K_n\,\phi\,\omega \quad .....(5.47)$$

Or

$$\omega = \frac{E_b}{K_n\phi} \quad .....(5.48)$$

In case of cumulative compound motor, the series field increases, the shunt main field flux. Hence as the load increases, the flux ϕ increases. So equation 5.48 indicates that the speed will decrease. Whereas for differential compound motor, the series field opposes the main shunt field. So, there is reduction in ϕ and equation 5.48 gives that the speed will increase. In Fig. 5.15 curves between speed ω and armature current I_a are given for a pure shunt motor, cumulative compound motor and differential compound motor.

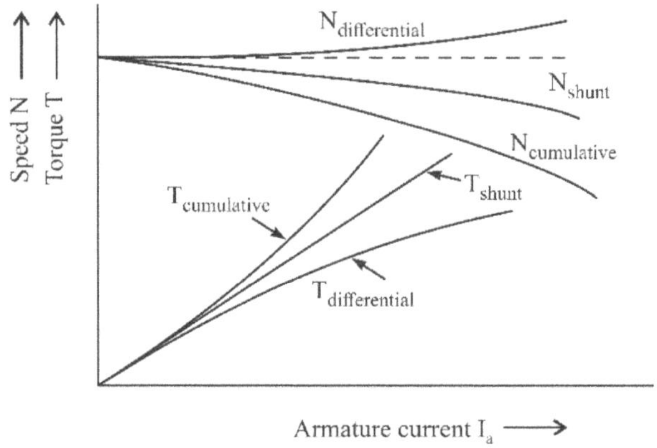

Figure 5.15 Characteristics of DC Compound Motor

It may be seen from Fig.5.15, that where high torque is required cumulative compound motor, is preferred and where stabilized speed is desired a differential compound motor, with fewer turns, is preferred.

With the help of speed ($N - I_a$) and torque ($T - I_a$), the speed torque ($N - T$) curves for shunt motor, cumulative and differential compound motors are shown in Fig. 5.16.

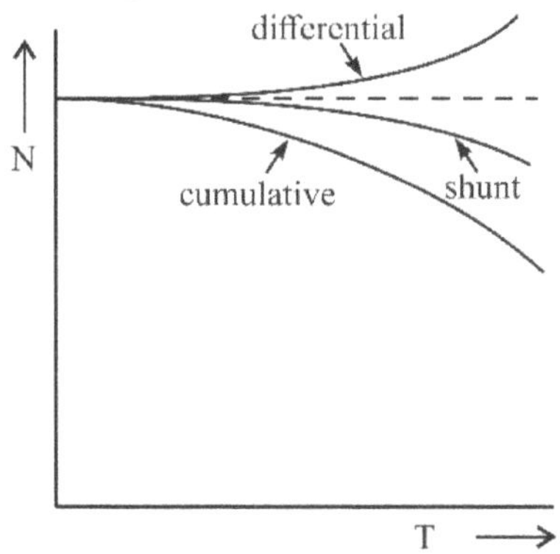

Figure 5.16 Speed torque characteristics of basic compound motor

5.6 Comparison of Different Types of DC Motors

For a proper comparison of characteristics of different types of DC motors, it has been considered that all the motors have same speed and same torque for same value of the full load current at rated supply voltage.

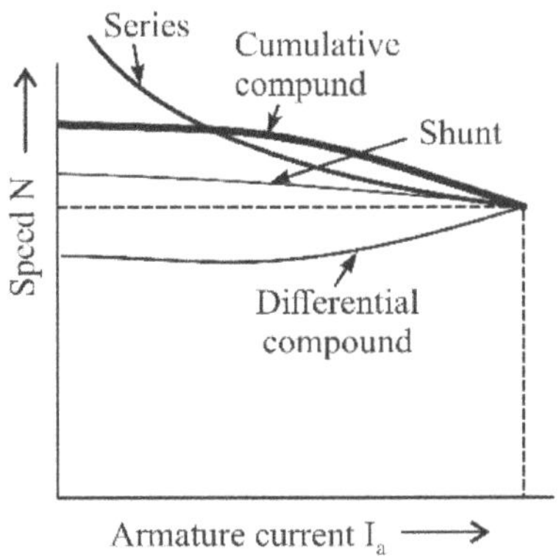

Figure 5.17 N-I_a Curves

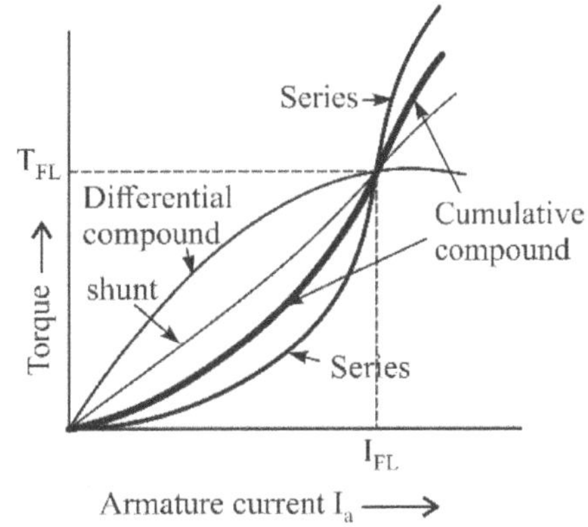

Figure 5.18 T-I_a Curves

The speed of a series motor increases by decreasing the load. At no load the speed tends to become infinite. This phenomenon is termed as Flying Away of the motor and the armature will wreck in to pieces. The armature conductor will suddenly come out of the armature slots.

The speed of the shunt motor decreases slightly from no load to full load. The decrease in speed is so small that it may be taken almost as a constant speed motor. The speed – armature current ($N - I_a$) characteristics for shunt, series, cumulative compound and differential compound are shown in Fig. 5.17. In cumulative compound motor the series field adds to the shunt field and the resultant field increases thereby the speed of the motor decreases with the increase in armature current. Whereas for differential compound motor the series field opposes the main shunt field. The resultant flux will decrease and the speed will increase by increasing the armature current.

In Fig.5.18 torque versus armature current ($T - I_a$) characteristics for shunt, series, cumulative compound and differential compound motors are given as explained earlier. The ($T - I_a$) curve of shunt motor is a straight line and for series motor it is hyperbolic till saturation. When magnetic circuit becomes saturated, the curve becomes linear. In cumulative compound motor the field increases with increasing current and differential compound motor field decreases with increasing current I_a. When flux increases speed decreases and when flux decreases speed increases. Fig.5.19 gives comparison of speed versus armature current ($N - I_a$) characteristics for various motors.

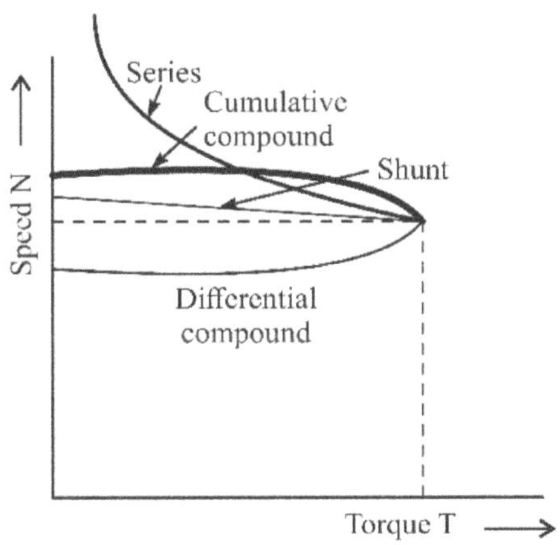

Figure 5.19 Speed –Torque characteristics

With the help of curves of Fig. 5.17 and 5.18, the curves between Torque T and Speed N are plotted for shunt, series, cumulative compound and differential compound motors in Fig.5.19.

5.7 Applications of DC Motors

A summary of characteristics and applications of different types of DC motors is given in the Table 5.1.

Table 5.1

Type of DC Motor	Characteristics	Applications
Series	(a) Very high starting torque (b) Varying speed, being dangerously high at no load (c) Self-adjustable varying speed at load.	(i) For traction work i.e., electric loco motive, trolley cars and buses. (ii) Cranes and hoists (iii) Conveyors (belt)
Shunt	(a) Nearly constant speed (b) Medium starting torque (1.5 to 2 times full load torque	(i) For driving constant speed shafts. (ii) Lathes, centrifugal pumps, reciprocating pumps. (iii) Fans and blowers (iv) Wood working machines, paper mills and printing presses.
Cumulative compound	(a) High starting torque (b) Varying speed but within limit (c) Self-adjustable varying speed at load.	(i) For intermittent high torque load (ii) Shears, punches and presses (iii) Elevators, conveyors and stokers (iv) Heavy rolling mills
Differential compound	(a) Constant speed but danger of running away (b) Low starting torque	Not generally preferred. Sometimes to maintain constant speed, a very few turns of differentially connected series field is used.

5.8 Necessity of DC Motor Starters

From equation 5.12 the back emf of the DC motor is given by $E_b = K_n \phi N$. At the time of starting, the motor armature is at rest and so there is no induced back emf.

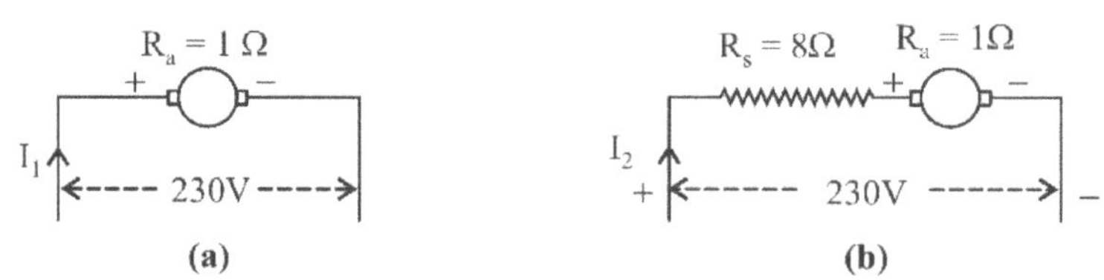

Figure 5.20 Circuit for example

Consider a 4 kW, 230 V DC motor having an armature resistance of 1 Ω. Taking 100% efficiency, the armature current at starting from Fig. 5.20 (a).

$$I_1 = \frac{230}{1} = 230 \text{ A}$$

Its full load current is

$$I_2 = \frac{4\times10^3}{230} = 17.4\ \text{A}$$

The motor is rated at 17.4 A of full load current, whereas the current drawn at starting $I_1 = 230$ A, which is extremely high current and will burn the armature conductors of the motor. But if a suitable value of resistance, say $8\,\Omega$, is put in series with the armature as shown in Fig. 5.20(b), the starting current I_2 will be

$$I_2 = \frac{230}{8+1} = 25.6\ \text{A}$$

Thus, the starting current is limited to 25.6 A, which is 1.47 times the full load current. This series resistance is termed as Starting Resistor. The value of the Starting Resistor is so selected that the starting current is 1.2 to 1.8 times the full load current.

5.8.1 Three-Point DC Shunt Motor Starters

In Fig.5.21 a three point shunt motor starter with No-Volt and Overload Relays is shown. The starter has three terminals marked as L, F and A. L indicates supply line terminal, F indicates field terminal and A the armature terminal. The starting resistor is divided into several steps and connected to terminal A.

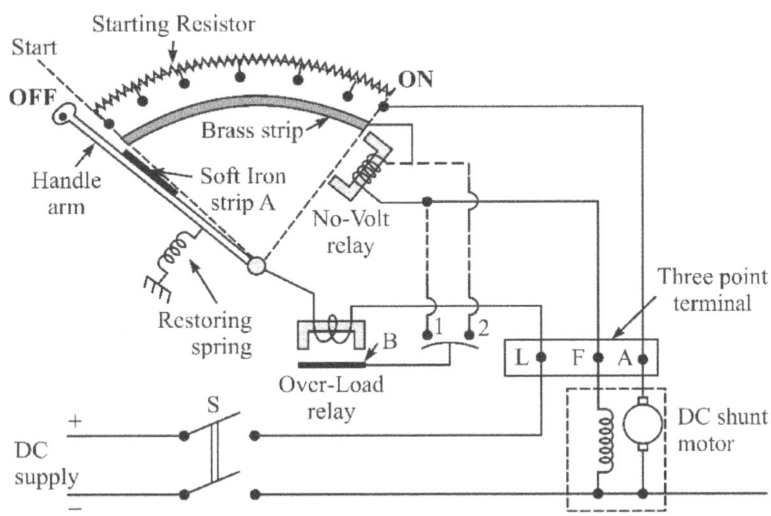

Figure 5.21 Three-point DC shunt motor starter

It may be noted that, full line current flows through the coil of Overload Relay and the shunt field current flows through the coil of No-Volt Relay. For starting the motor, the switch S is made on. No current flows through the motor as the handle is in OFF position. Now the handle is moved to the first stud. Immediately, circuits of armature and shunt field get energized. For armature circuit the current flows through the magnetizing coil of Overload Relay, handle arm, starting resistor and the armature, returning to -ve of the supply line. For the shunt field circuit, the current flows through the handle arm, brass strip, coil of No volt relay and shunt field returning to -ve of the supply line. Since both armature and field circuits get energized, the motor starts running and develops back emf. The handle arm is moved to the next step stud, the motor further picks up steed. The handle arm is moved to next step stud giving a time of 2 to 3 sec for picking up speed till it reaches the last stud. Now the starting resistor is fully removed from the armature circuit and the motor achieves its full speed. The function of No-Volt Relay and Overload Relay is described below.

No-Volt Relay

It consists of a soft iron strip A attached to the handle arm, and an electro magnet. The magnetizing coil of the electro magnet is in series with shunt field winding. The same shunt field current is flowing through this coil and the No-Volt relay gets magnetized attracting the soft iron strip A and keeping in ON position. If the supply goes, there is no more current flowing through this coil and the No-Volt Relay gets demagnetized and the handle arm returns to OFF position through the action of Restoring spring. Its advantage is that when the supply comes again, the starter handle is in OFF position and no current will flow to the motor. It can be safely restarted. If the handle is allowed to remain in the ON position, there would not be any starting resistor in series with armature resulting in burning of the motor.

Overload Relay

It also consists of a soft iron piece B and an electro magnet. Through the magnetizing coil of the Overload Relay the full line current always flows so long as the motor is running. The electro magnet of the Overload Relay is designed in such a way that it cannot attract the soft iron piece B up to the full load current. When the motor is overloaded say by 20% or so, the input current becomes greater than the full load current, B gets attracted and short circuits the points 1 and 2 resulting in short circuiting of coil of the No-Volt Relay. This is achieved by connecting the ends of coil of No-Volt Relay to two brass terminals 1 and 2 as shown by dotted lines in Fig.5.21. In this case no current flows through the coil of No-Volt Relay and it gets demagnetized, relaxing the handle arm to return to OFF position. The motor is automatically protected against overloading of the motor.

While designing the Overload Relay, it is normally preset for 15%, 20% or 25% overload current. Thus, by providing No-Volt Relay and Overload Relay, the motor is protected both against No supply and overload.

Precaution

The handle arm should be moved slowly from one stud to next stud to allow time for picking up of speed. If the handle is moved suddenly from OFF position to ON position, there would be no time to pick up speed to develop back emf resulting in likely burning of the motor and starter.

Taking the case of machines in the laboratories, the order of resistance of magnetizing coil of overload relay is $0.2 - 0.5\,\Omega$ and of the magnetizing coil of No Volt Relay $200\text{-}400\,\Omega$.

5.8.2 Four-Point DC Shunt Motor Starters

In three-point starter, the magnetizing coil of the No-Volt Relay is in series with the field winding. The resistance of this coil is quite high, so the total resistance in the shunt field circuit becomes very high reducing the shunt field current. The developed torque is proportional to field current, which, if low, will develop low starting torque. Further, in No-Volt Relay circuit, the low value of field current through its may disrupt the motor working. Sometimes the handle may jump back to OFF position during normal running due to weak magnetization of the No-Volt Relay. To overcome this problem, a 4-point starter is used.

In this case the field current directly goes to shunt field winding and No-Volt Relay coil is in series with a Safety Resistor is directly connected across the supply. The complete connection diagram of 4-point shunt motor starter is shown in Fig.5.22. To start the motor the Handle is brought to START position. There are three paths for the current.

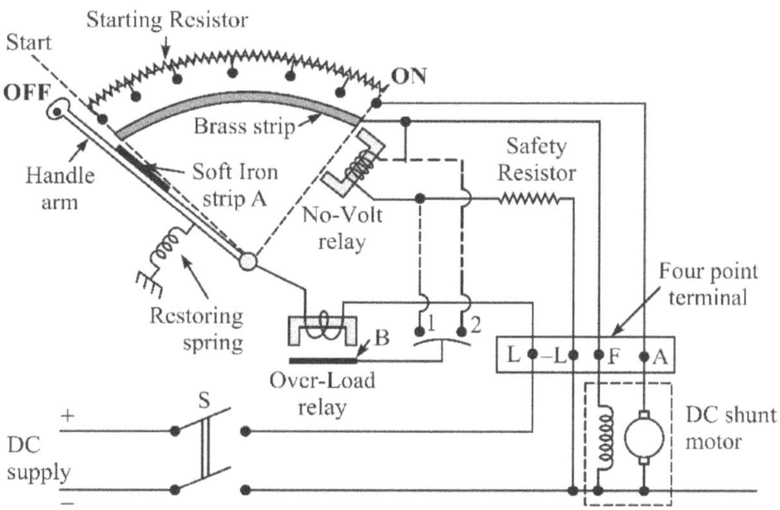

Figure 5.22 Four-point DC Shunt Motor Starter

(i) From +ve line current flows to Starting Resistor through Overload Relay coil, the handle arm to armature of the motor and returns to the -ve terminal.

(ii) The second path of current from Handle arm is to Brass strip to field winding of the motor returning to -ve terminal.

(iii) The third path of the current from Brass strip is to magnetizing coil of the No-Volt Relay to the Safety Resistor returning to -ve terminal.

As the motor current is flowing through the motor armature and shunt field, torque will be developed and rotor starts running. The motor speeds up and the handle moved from one stud to the next, till the entire Starting Resistor is cut off. The handle arm is now in the "RUN" position.

Since the current is flowing through the coil of the No-Volt Relay, it is sufficiently magnetized and holds the soft iron strip of Handle arm in the RUN position. If there is no supply, the No-Volt Relay gets demagnetized and the Handle arm returns to the OFF position. If there is an overload, greater or equal to the pre-set value, Overload Relay shall be sufficiently magnetized to short circuit terminals 1-2, short circuiting the magnetizing coil of the No-Volt Relay. The current will not flow through the coil and will release the handle arm to the OFF position. The path of current shall be from the Brass strip to shorted points 2-1 through the safety Resistor to the -ve terminal. If there is no safety resistor, line terminals +ve and -ve will get shorted. Hence a Safety Resistor of suitable value is very much essential. Note that the number of studs in a starter is equal to the number of steps or sections plus 1.

5.8.3 DC Series Motor Starters

Series motor starter consists of a Graded Series Resistor inserted in the armature circuit. The graded resistor has smaller sectional area in the beginning, which increases till the end of the resistor. The idea is that as the motor is started, the earlier part of the resistor is in the circuit for a shorter period and the last part for longer period. Hence the current carrying capacity of the last part should be higher.

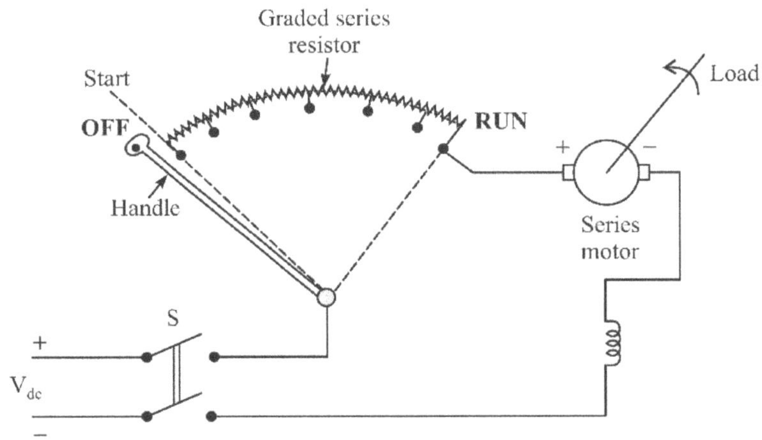

Figure 5.23 Starting of series motor

A simple circuit diagram of starting of series motor without No-Volt and Overload Relays is shown in Fig. 5.23.

For starting the motor, the supply switch S is made ON and the handle is moved to start position. The current flows through the Graded Resistor, armature and series field of the motor. Series motor should not be started without load as there is a danger of flying away. The motor shall start running. The Graded Resistor is removed slowly from the circuit in steps. For speed control purpose, some part of the resistor may be in the circuit, but there is heavy loss of energy in the resistor. For practical purposes Drum Controller are employed for starting as well as speed control.

5.8.4 Calculation of Step Resistors for DC Shunt Motor Starters

Fig. 5.24 gives representation of n step shunt motor starter.

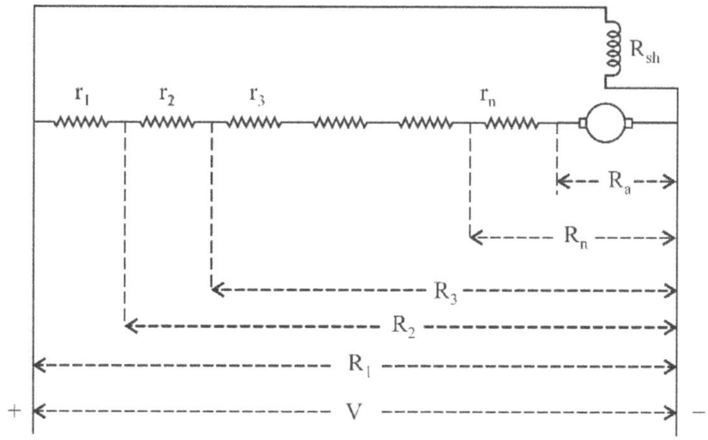

Figure 5.24 Step resistors in a starter

The step resistors are r_1, r_2, r_3 .... R_n and supply voltage in V. When supply voltage V is applied, the motor is stand still, the current through the armature is I_1 given by

$$I_1 = \frac{V}{R_1} \quad \text{.....(i)}$$

As armature and field are both energized, the motor starts running and develops back emf E_{b1}. The current drops to I_2, given by

$$I_2 = \frac{V - E_{b_1}}{R_1} \quad \text{.....(ii)}$$

I_2 is taken as full load current of the motor and I_1 is the starting current which is normally 1.2 to 1.8 times the full load current I_2. The ratio I_1/I_2 may be taken equal to x (x being 1.2 to 1.8). The values of r_1, r_2, r_3 .... are so selected that as we move from one step to another. Current rises to I_1, and on development of back emf it drops to I_2. At this point the resistor r_1 is cut off, so that the resistance in the armature circuit R_2. The current will at once rise to I_1 given by the relation

$$I_1 = \frac{V - E_{b_2}}{R_2} \quad \text{.....(iii)}$$

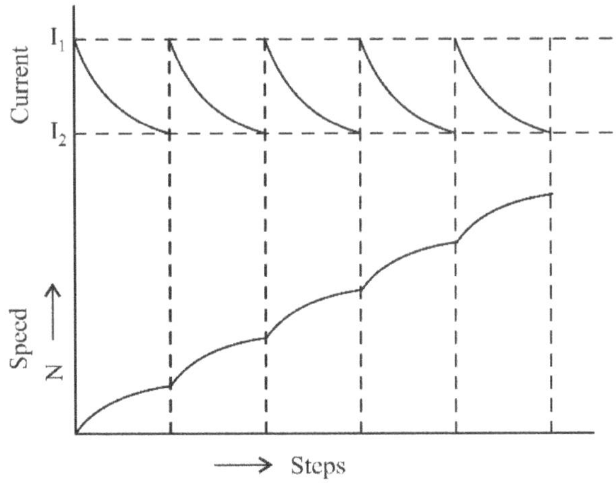

Figure 5.25 Speed and Current in starter

The motor speeds up and develop back emf E_{b_2} and resistance in the circuit remaining R_2. The current drops to I_2 given by relation.

$$I_2 = \frac{V - E_{b_2}}{R_2} \qquad \text{.....(iv)}$$

Now the resistance r_2 is cut off, so that the resistance in the armature circuit is R_3. The armature current shall jump to I_1 given by

$$I_1 = \frac{V - E_{b_2}}{R_3} \qquad \text{....(v)}$$

We proceed likewise and cut off all step resistors till the last resistor r_n is in the armature circuit developing the back emf E_{b_n} giving the relation

$$I_2 = \frac{V - E_{b_n}}{R_n} \qquad \text{.....(vi)}$$

Finally, the resistor r_n is now cut off. Only armature resistance R_a remains in the circuit. The current will jump to I_1 given by

$$I_1 = \frac{V - E_{b_n}}{R_a} \qquad \text{.....(n+1)}$$

Divide equation (iii) by (ii), we get

$$\frac{I_1}{I_2} = \frac{R_2}{R_3} \qquad \text{.....(n + 2)}$$

Divide equation (v) by (iv)

$$\frac{I_1}{I_2} = \frac{R_2}{R_3} \qquad \text{.....(n+3)}$$

and so, on

Divide equation (n + 1) by (n)

$$\frac{I_1}{I_2} = \frac{R_n}{R_a} \qquad \text{.....(n + m)}$$

Multiplying all the equations (n + 2), (n + 3) …. (n + m), we get

$$\left(\frac{I_1}{I_2}\right)^n = \frac{R_1}{R_2} \times \frac{R_2}{R_3} \times \frac{R_3}{R_4} \; \ldots\ldots\ldots \; \frac{R_n}{R_a}$$

$$= \frac{R_1}{R_a}$$

$$\therefore \qquad x = \frac{I_1}{I_2} = \sqrt[n]{\frac{R_1}{R_a}} \qquad \text{.....(n + m + 1)}$$

If we know x, the ratio of starting to full load current, (I_1/I_2), and the number of steps in starter, we can find R_1.

From equation (n + 2)

$$R_2 = \frac{R_1}{I_1/I_2} = \frac{R_1}{x}$$

From equation (n + 3)

$$R_3 = \frac{R_2}{x}$$

Thus, $r_1 = R_1 - R_2$

$r_2 = R_2 - R_3$

$r_3 = R_3 - R_4$

Thus, all resistance can be easily calculated. All the step resistors can be easily calculated. Fig.5.25 shows rise of current I_1, and then drop to I_2, due to development of back emf. On each step current jumps to I_1 and then drops to I_2 in exponential way and the speed N rises as shown.

Example 5.5: A 230 V shunt motor has an armature resistance of 0.4 Ω. The starter has 6 steps. If the maximum armature current is 50 A, determine the resistances of the successive steps and the lower current limit.

Solution: Supply voltage V = 230 V

Maximum current I_1 = 50 A

Number of steps n = 6

Armature Resistance $R_a = 0.4\ \Omega$

The lower current limit I_2 is normally the rated current of the motor.

Total resistance R_1 at starting is:

$$R_1 = \frac{\text{Applied voltage V}}{\text{Starting current (Maximum current } I_1\text{)}}$$

$$= \frac{230}{50} = 4.6\ \Omega$$

In a starter, the important relation is

$$x = \frac{I_1}{I_2} = \sqrt[n]{\frac{R_1}{R_a}}$$

Or $$x = \frac{50}{I_2} = \sqrt[6]{\frac{4.6}{0.4}} = 1.5$$

& $$I_2 = \frac{I_1}{x} = \frac{50}{1.5} = 33.3\ \text{A}$$

Here $$R_2 = \frac{R_1}{x} = \frac{4.6}{1.5} = 3.07\ \Omega$$

$$R_3 = \frac{R_2}{x} = \frac{3.07}{1.5} = 2.04\ \Omega$$

$$R_4 = \frac{R_3}{x} = \frac{2.04}{1.5} = 1.36\ \Omega$$

$$R_5 = \frac{R_4}{x} = \frac{1.36}{1.5} = 0.91\ \Omega$$

$$R_6 = \frac{R_5}{x} = \frac{0.91}{1.5} = 0.61\ \Omega$$

The step resistances are given by

$r_1 = R_1 - R_2 = 4.6 - 3.07 = 1.53\ \Omega$

$r_2 = R_2 - R_3 = 3.07 - 2.04 = 1.03\ \Omega$

$r_3 = R_3 - R_4 = 2.04 - 1.36 = 0.68\ \Omega$

$r_4 = R_4 - R_5 = 1.36 - 0.91 = 0.45\ \Omega$

$r_5 = R_5 - R_6 = 0.91 - 0.61 = 0.30\ \Omega$

$r_6 = R_6 - R_a = 0.61 - 0.40 = 0.21\ \Omega$

Here the number of starter studs

= Number of steps + 1

= 6 + 1 = 7

5.9 Introduction to Soft Starting of DC Shunt and Series Motors

By soft starting of motors it is understood regarding development of solid state control circuits to give reduced voltage suitable for starting of the DC motors. In normal resistor based starting a series resistor is placed in series with the armature so that the available voltage to the armature make a current of about one and a half times of the full load current. The resistors are cut in steps and then full voltage is applied and motor picks up its full speed. Where as in soft starting, reduced voltage is obtained from a solid state control circuit in two ways:

1. By using Controlled rectifier circuits usually named as Converters.
2. By using Chopper circuits in which ON and OFF time of the thyristor is changed.

5.9.1 Controlled Rectifier Circuit

The circuit arrangement in Fig. 5.26.

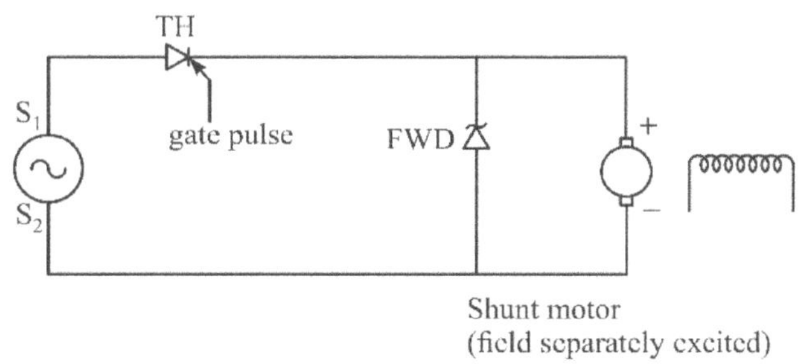

Figure 5.26 Starting of Shunt motor (Field separately excited)

The simplest converter with 1 pulse per cycle is shown in Fig.5.27. During stand still conditions of the DC motor, when S_1 is more positive than S_2, the thyristor TH may be turned on by giving a gating pulse to the gate of the thyristor, producing output voltage. Let the pulse be applied at firing angle α, the rectified voltage is shown by hatched area in one cycle. During the -ve half cycle, there would be no rectification. As the motor circuit is inductive in nature, a Free Wheeling Diode (FWD) is required for dissipation of inductive energy by providing an alternative current path.

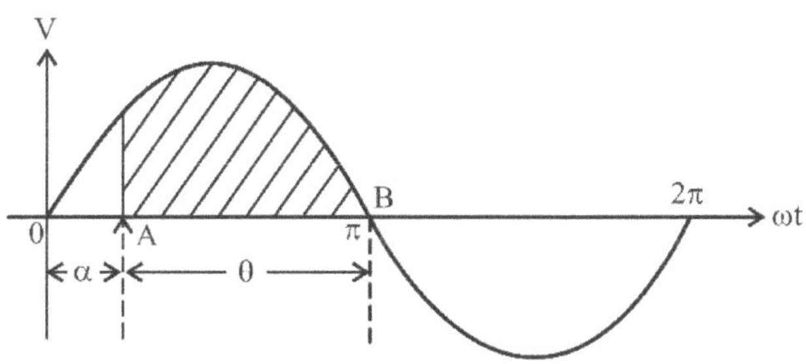

Figure 5.27 Rectified Voltage

The applied sinusoidal voltage to the circuit of Fig. 5.26 be given by

$$e(t) = E_m \sin \omega t \quad \text{.....(i)}$$

Average voltage obtained is

$$E_{av} = \frac{1}{2\pi}\int_{\alpha}^{\pi}(E_m \sin\theta)d\theta \quad \text{.....(ii)}$$

E_m is the peak value of the single-phase ac applied voltage, α is firing angle and θ (= ωt) is varying from zero to π in half cycle, when conduction is taking place
equation (ii) gives

$$E_{av} = \frac{1}{2\pi}[-E_m \cos\theta]_{\alpha}^{\pi} = \frac{E_m}{2\pi}(1+\cos\alpha) \quad \text{.....(iii)}$$

For firing angle $\alpha = 0$, $E_{av} = \dfrac{E_m}{\pi}$

For firing angle $\alpha = \pi$, $E_{av} = 0$

For firing angle $\alpha = \pi$, the output voltage is zero. Thus, by adjusting firing angle α, the desired dc output voltage which may be obtained which may be suitable to start the motor. In the circuit shown in Fig.5.26, the field is separately excited.

For full wave-controlled rectification the circuit is shown in Fig. 5.28.

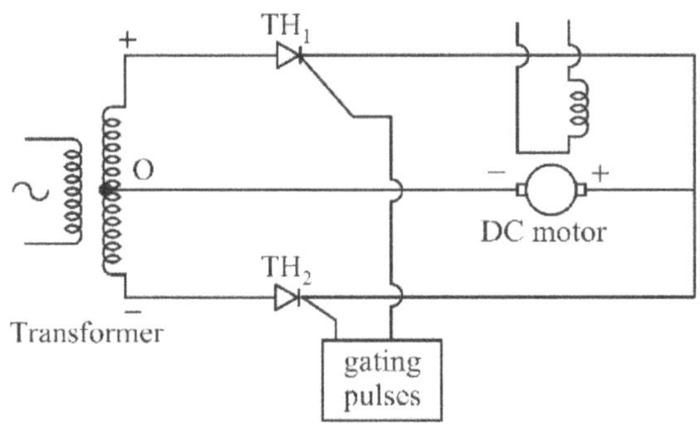

Figure 5.28 Full wave controlled rectifier for speed control

The supply is taken from a centrally tapped transformer. Two Thyristors are used for full wave rectification. When one terminal of the supply transformer is +ve, the other is -ve. The thyristor TH_1 is fired at angle α. The rectified voltage, as shown by hatched area, will be available as shown in Fig. 5.29. The current will flow through TH1 to the +ve of DC motor armature and to the mid-point "O" of the transformer. In the next half cycle the lower terminal of the secondary will be +ve and current will flow through TH_2 to the +ve terminal of the armature and return to mid-point "O" of the secondary of the transformer. Thus, rectification is obtained in the other half cycle of the supply voltage as shown in Fig.5.29.

The average voltage now obtained is

$$E_{av} = \frac{1}{\pi}\int_{\alpha}^{\pi}(E_m \sin\theta)d\theta$$

$$= \frac{1}{\pi}\left[-E_m \cos\theta\right]_{\alpha}^{\pi}$$

$$= \frac{E_m}{\pi}(1+\cos\alpha) \qquad \text{.....(iv)}$$

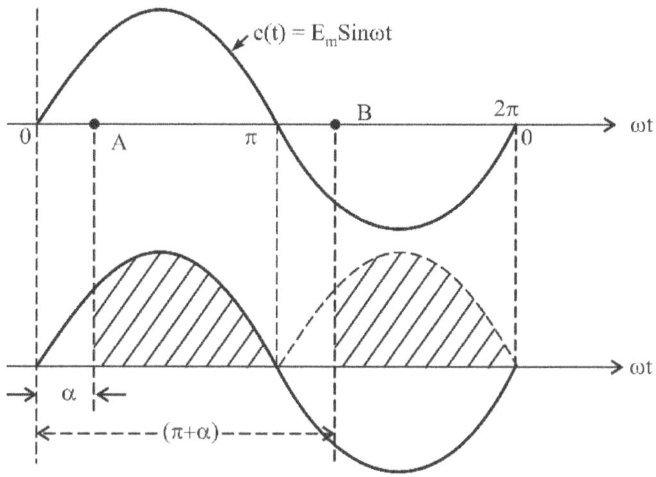

Figure 5.29 Rectification in both halves

Comparison of equations (iii) and (iv) show that average voltage with full wave rectification is double to that of the half wave rectification in one cycle.

5.9.2 Chopper Circuit

A simple chopper circuit to start a DC series motor at reduced voltage and then to run it at the full voltage is given in Fig. 5.30.

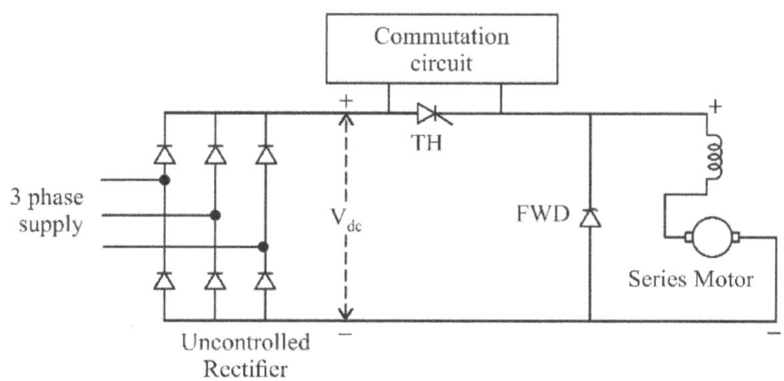

Figure 5.30 Chopper circuit with DC series motor

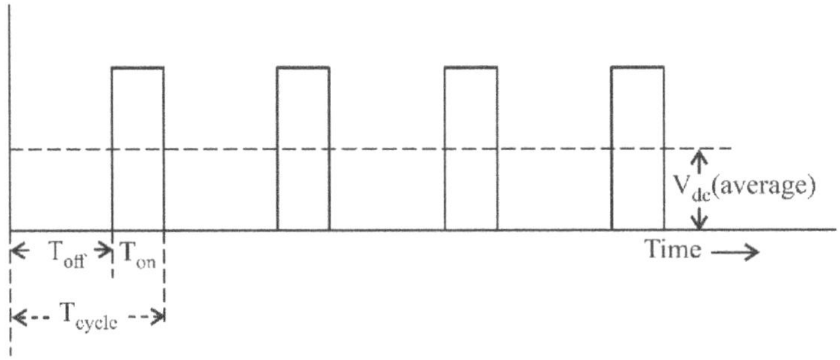

Figure 5.31 Wave shape of chopper output voltage

In circuit of Fig.5.30, V_{dc} is obtained from three phase ac supply using uncontrolled diode bridge rectifier and regulating the output voltage by means of dc chopper.

In the series chopper circuit, a thyristor TH is placed in series with the dc supply. It is rapidly switched on and off to obtain controlled dc output voltage consisting of a series of rectangular voltage pulses as shown in Fig.5.31.

Average dc voltage is given by

$$V_{dc(average)} = \frac{T_{on}}{T_{cycle}} V_{dc} = \frac{T_{on}}{T_{on} + T_{off}} . V_{dc} \qquad .....(v)$$

T_{on} is ON time when thyristor TH is conducting and T_{off} when TH is not conducting.

By controlling ON and OFF times, the average dc voltage is easily controlled from very low voltage to full voltage. This type of starting is termed as soft starting of DC motors.

5.10 Speed Control of DC Motors

In order to learn about the speed control of DC motors, it is required to recall the following equations.

$V_t = E_b + I_a R$

where $R = R_a$ in case of shunt motors and $R = (R_a + R_f)$ in case of series motors

Also, $$E_b = \frac{P\phi NZ}{60a}$$

or $E_b \propto \phi N$

Any motor, running at its rated speed with rated terminal voltage, delivers rated electromagnetic torque (T_e) to the load. Under such a condition $T_e = T_L$ and the motor keeps running at constant speed.

In this section of speed control, it is assumed that V_t and T_L remain constant while the speed of the motor is being controlled. The expression for speed is derived from the above two equations as

$$N \propto \frac{V_t - I_a R}{\phi}$$

It can be seen that the speed can be varied by varying

(i) the flux ϕ, called the Field Control method

(ii) the I_a R drop, called the Armature control method; and

(iii) the V_t control called Ward Leonard method or Thyristor voltage control.

5.10.1 Field Control Method

The magnetic flux is controlled in this method by inserting a rheostat (Rh) in the field circuit of the motor.

5.10.1(a) DC Shunt Motor

In Field control method DC shunt motor, a rheostat is inserted in series with the field winding as shown in Fig 5.32 (a) to control the field flux. Initially, when the value of resistance provided by the rheostat is zero and the motor is running at its rated (i.e. full load) speed, the load torque T_L (exerted by the load) is equal to the electromagnetic torque T_e (generated by the motor). Thus, $T_e = T_L$.

Now when a resistance is inserted in the field circuit (in series with the field winding) through the rheostat, the overall resistance of the field circuit increases and with the applied voltage being constant, there happens to be a reduction in the field current.

As the field current is proportional to the field flux (in the linear region), there happens to be a reduction in the field flux.

The back emf $E_b = K_n \phi N$ gets reduced because of the reduction in field current (there can be no instantaneous change in motor speed due to the moment of inertia of the rotor).

The reduction in E_b causes an increase in the armature current I_a; as per the formula $I_a = (V - E_b)/R_a$

Now the electromagnetic torque $T_e = K_t \phi I_a$, has field flux getting reduced and the armature current getting increased. But the amount of increase in the armature current is several times more than the amount of reduction in the field flux.

Hence there is an overall increase in the electromagnetic torque T_e.

The increase in T_e (with the load, and hence T_L remaining constant) causes the motor to accelerate and hence its speed increases. The characteristic depicting rise in speed with weakening field is shown in Fig 5.32 (b).

It is thus concluded that only the speeds higher than the rated speeds can be obtained using this method of speed control. Further, this method can be termed as a constant power method, wherein the torque gets reduced, while the speed gets increased, maintaining the same power. This is shown in Fig 5.32 (c).

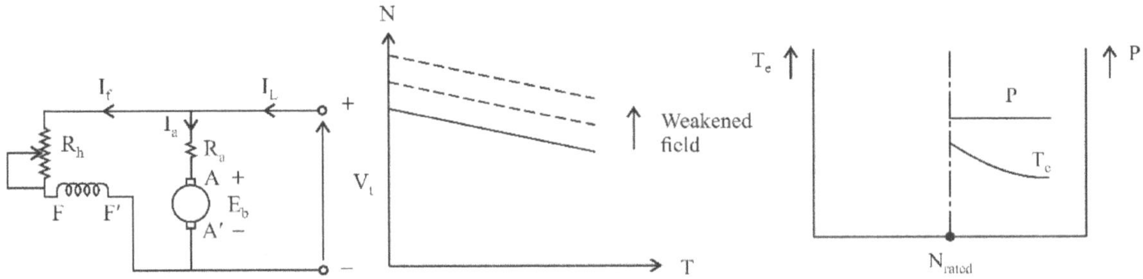

Figure(a) Circuit **Figure(b)** (N – T) characteristic **Figure(c)** (N – T) and (N – P)

Figure 5.32 Field control of DC shunt motor

5.10.1(b) Field Control of DC Series Motor

The field current of DC series motor can be controlled by various methods such as; by inserting a diverter rheostat, by having tappings in the field winding, by connecting partial field windings in series-parallel. The arrangements are shown in Fig.5.33. The terminal voltage is maintained constant.

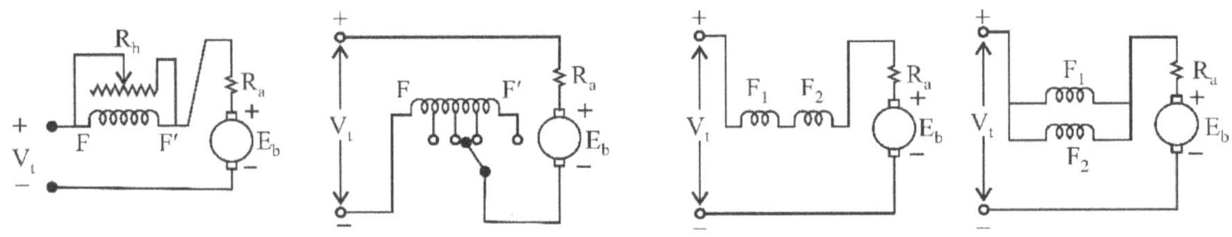

(a) Field Diverter (b) Tap changer (c) Series- Parallel connection

Figure 5.33 Field control of DC series motor

The field diverter, as shown in Fig. 5.33(a), facilitates diversion of the field current through an external rheostat Rh. Any fraction of field current can be obtained to flow through the field winding. An inductive resistance is normally used as otherwise smooth speed control may not be achieved.

Use of tap changer, as shown in Fig.5.33(b), gives higher speeds by reducing the field winding turns. This arrangement is normally used in traction. In series parallel arrangement of field control, as shown in Fig.5.33 (c), higher speed is obtained when the two parts of field winding are connected in parallel.

Thus, both in shunt and series DC motors when the speed is controlled by Field control method, maintaining constant terminal voltage, the nature of speed-torque characteristics remains nearly the same.

5.10.2 Armature Control Method

In this method of speed control an external rheostat is inserted in the armature circuit of the motor. This method is also known as Rheostatic method of speed control. Typically, the armature current is high in motors, hence use of external resistance in armature circuit causes lot of I^2R loss in it.

Also, as against field control method this method provides only the speeds lower than the rated speed.

5.10.2(a) Armature Control of DC Shunt Motor

In the armature control method of speed control of a DC shunt motor, a rheostat is connected in series with the armature circuit as shown in Fig 5.34 (a).

Initially, when the resistance offered by the rheostat is zero ohm, the motor is running at its rated speed.

When the value of rheostat resistance is increased, with the applied voltage constant, the fist natural reaction is the reduction of the armature current.

$$I_a = (V - E_b)/R_a$$

And with resistance introduced through the rheostat, the armature current will be reduced to

$$I'_a = (V - E_b)/(R_a + R_{ext})$$

where R_{ext} is the rheostat resistance.

With this lower armature current, the torque

$$T_e (= K_t \phi I_a) \text{ gets reduced to } T_e' (= K_t \phi I_a')$$

Now since the load torque T_L is unchanged, and the driving electromagnetic torque T_e is reduced, the speed also gets reduced. The reduction in speed corresponding to increase in resistance of Rheostat is shown in Fig 5.34 (b).

This reduction in speed subsequently causes a reduction in the back emf,

$$E_b = K_n \phi N;$$

which, in turn, causes an increase in the armature current back to its original value.

$$I_a = (V - E_b)/R_a = (V - E_b)/(R_a + R_{ext})$$

Thus, we may conclude that with the armature current getting back to its original value and flux remaining unchanged, the electromagnetic torque T_e gets back to its original value and it can thus be concluded that this method is also termed as the constant torque methods, wherein the motor continues to run at a reduced speed feeding the same load torque T_L.

It can also be said that only the speeds, below the base (or rated) value, can be achieved through this method.

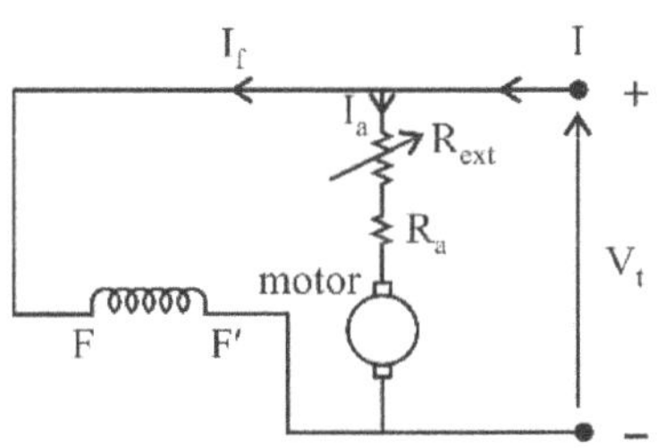

Figure(a) Circuit arrangement

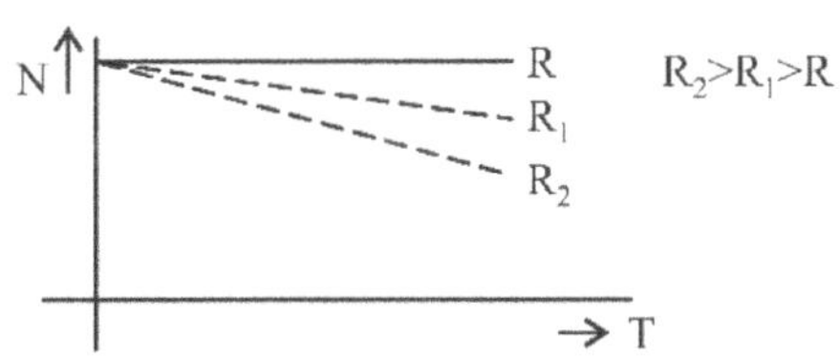

Figure(b) Speed-torque characteristics

Figure 5.34 Armature control of DC shunt motor

To summarize the discussion so far it is established that the speed can only be increased by using field control method and can only be reduced by using armature control method. The speed is increased nearly up to 25% of rated value due to sparking at brushes; and the speed is reduced usually up to 30% to 50% of its rated value. Much of power loss takes place in armature control as compared to field control. Speed regulation is better in field control. The variation in power and torque with speed are shown in Fig. 5.35.

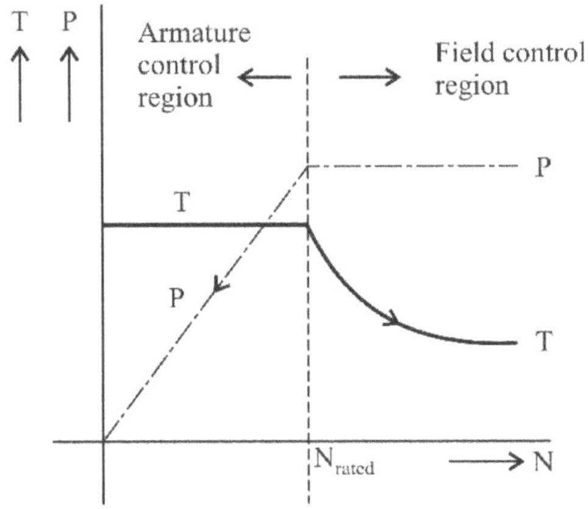

Figure 5.35 Constant torque and constant power operation characteristics

5.10.3 Ward Leonard Method (Voltage Control)

This is an expensive method of speed control as it requires an additional Motor – Generator set. The arrangement of machines is as shown in Fig.5.36.

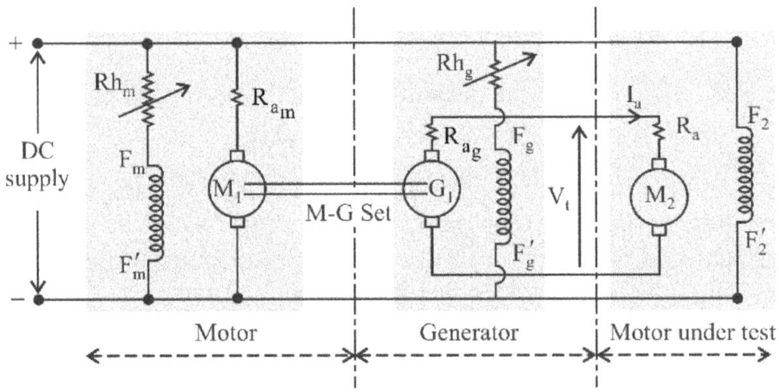

Figure 5.36 Ward Leonard method of speed control

The circuit arrangement of Fig.5.36 shows connections of 3 DC shunt machines. Machines M_1 and G_1 form an M – G set having field control rheostats connected to them. The aim is to control the speed of the motor M_2, the Motor Under Test, by regulating its supply voltage to armature.

Here, the speed of motor M_1 (using Rh_m) is adjusted equal to the rated speed of generator G_1 (say N_g). Machines M_1 and G_1 are mechanically coupled together. For generator G_1 we know that

$$N_g \propto \frac{V_t + I_a R_{ag}}{\phi_g}\text{; which is maintained constant}$$

The value of V_t and can be changed by changing the value of ϕ_g by using Rh_g. This V_t is the supply voltage in the motor M_2. Hence the speed of motor (N_m) gets changed as follows.

$$N_m \propto \frac{V_t - I_a R_{am}}{\phi_m}\text{, where } \phi_m \text{ is constant}$$

It is clear from the above discussion that since ϕ_g can only be reduced, leading to reduction in V_t (as N_g has to remain constant). Therefore, speeds lower than rated speeds only can be obtained for M_2. However, speeds above rated speed may be obtained by usual field control method (by having a rheostat in M_2 field circuit as well). Additionally, one can reverse the rotation of M_2 by having an arrangement to reverse V_t across its armature terminals of M_2.

Though this method is very costly having poor efficiency, but it gives accurate speeds with good speed regulation. This arrangement finds application in large reversing rolling mills.

5.11 Losses and Efficiency

The main losses in DC motors are classified as follows.

1. Loss in magnetic circuit – Iron loss
2. Loss in electrical circuit – copper loss (Variable)
3. Mechanical loss

5.11.1 Iron Loss

The main dc field flux distribution in air gap gets distorted due to armature reaction. In case of motors the flux gets enhanced at the entering side of the pole shoes. With the increase in load current, the iron losses at teeth ends increased. Further, a magnetic saturation may occur at teeth ends, resulting in increased flux through core end – plates. This causes increase in iron loss there.

In other words, flux swings around pole shoe end (at entering end in motors and at leaving end in generators). This pulsating swing of flux depends on the speed with which the teeth cross a pole shoe one after the other. It depends upon the number of teeth multiplied by the armature speed in rps. The swinging and pulsating flux at pole – shoes end gives rise to eddy current and hysteresis losses. This iron loss depends on armature reaction and hence on armature current. The full load iron loss is nearly 50% more than that at no load condition.

The iron loss, described above, can be maximized by using laminated poles shoes or laminated poles. Also the ratio of the width of slot to the width of tooth can be kept low.

5.11.2 Copper Loss

Copper losses are the I^2R losses occurring in all the windings in a DC machine, such as in Armature winding, Field Winding, Interpole Winding, Compensating Winding. In addition to these, a current dependent brush contact loss (=Number of brushes × Load current × Voltage drop across brush) also takes place. Copper loss also takes place across the external resistance used in either armature or field circuit to facilitate speed control.

5.11.3 Mechanical Loss

Mechanical losses take place in the form of bearing friction, brush friction and windage.

5.11.4 Efficiency

The efficiency of a DC motor is defined as the mechanical power output to the electrical power input.

$$\text{Efficiency} = \frac{\text{Output}}{\text{Input}} = \frac{\text{Input} - \text{loss}}{\text{Input}}$$

The losses in a DC motor are categorized as:

1. Constant loss – No load loss, friction and windage loss (a).
2. Loss proportional to armature current – Brush contact loss (bI).
3. Loss proportional to square of armature current. All copper losses and additional iron loss at pole shoes (cI^2), therefore

The expression for efficiency can be given as

$$\eta = \frac{VI - P_L}{VI} = 1 - \frac{a + bI + cI^2}{VI}$$

$$= 1 - \left(\frac{a/V}{I} + \frac{b/V}{I} + \frac{c/V \cdot I}{1} \right)$$

$$= 1 - \left(\frac{a'}{I} + b' + c'I \right)$$

From above, the efficiency curve can be derived as shown in Fig.5.37

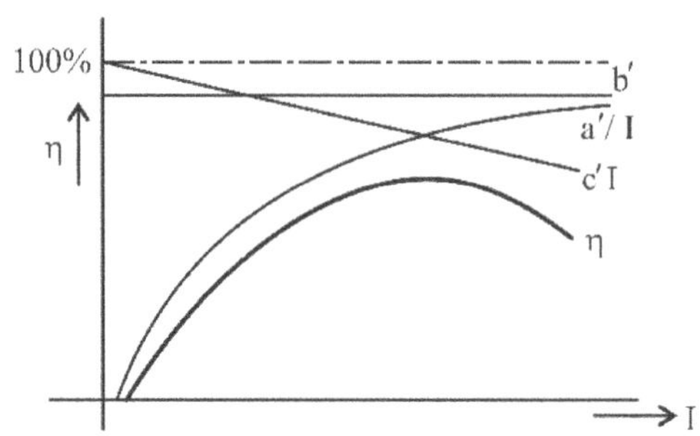

Figure 5.37 Efficiency curves

In order to obtain the condition for maximum efficiency, the losses (P_L) should be minimum. For this, its derivative $\left(\frac{dP_L}{dI} \right)$ should be equal to zero. So,

$$P_L = \left(\frac{a'}{I} + b' + c'I \right)$$

$$\frac{dP_L}{dI} = \left(a' + 0 + c'I^2\right) = 0$$

Or $a' = c'I^2$

That is, the efficiency will be maximum when the constant loss of the motor is equal to the losses varying with square of current.

5.12 Tests on DC Motors

Tests on DC machines are performed to know their losses, to find their efficiency and to observe the heat generated in them. The electrical input to DC motors (and electrical output of DC generators) can be measured using appropriate meters. However, the mechanical output of DC motors (and mechanical input to DC generators) are calculated by measurement of speed and torque. The electrical power (P_e) = $V_t \times I_L$, and the mechanical power (P_m) = $\frac{2\pi}{60}NT$.

The DC motors are required to be loaded while performing tests on them. The classification based on loading is as follows.

1. *Direct loading*: Load is directly applied to the motor using pulley – belt arrangement for brake test.
2. *Indirect loading*: Much energy gets lost in a direct loading method. To prevent this, the motor is tested by indirectly loading the motor. Swinburne's test and Hopkinson's test fall under this category.

5.12.1 Brake Test (Direct Method)

This test is performed on DC motors of smaller size. For motor of around 2 hp and above, water cooling is required to dissipate the heat generated due to friction. Arrangement for brake test is shown in Fig. 5.38.

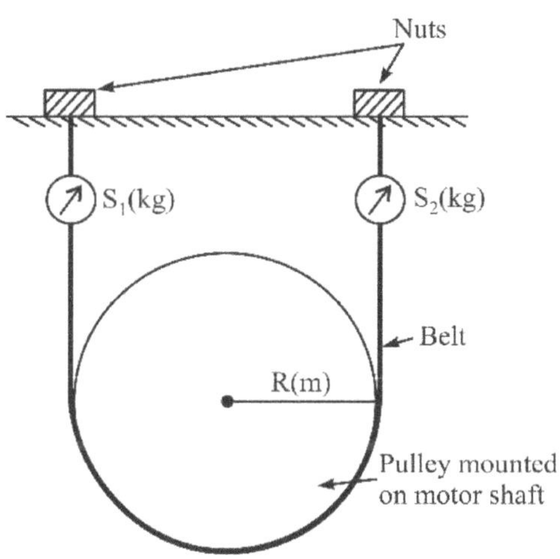

Figure 5.38 Brake test

The belt is tightened with the help of nuts. The belt applies braking effect on motor. The brake is applied till the motor draws its rated current (I_L) at rated terminal voltage (V_t). The readings of spring balances are noted as S_1 and S_2.

The mechanical load $T = [9.81 \times (S_1 \sim S_2) \times (R + r)]$ nm

where r is the thickness of the belt

$$\text{The output power} = \frac{2\pi NT}{60} \text{ watts,}$$

where N is the rpm of motor

The input power = $V_t\, I_L$ watts

Losses = Input power – output power

Efficiency = Output power/Input power

This method is used for smaller size motors only because it becomes difficult to dissipate the heat generated due to brake. The arrangement becomes bulky and its cooling becomes difficult for large motors. Readings of spring gauges keep fluctuating. In case of DC series motor, it is necessary to start the motor only after applying the brake beforehand.

5.12.2 Swinburne's Test (Indirect Method)

In order to find the losses and efficiency of larger DC machines, Swinburne's test is performed. This test is categorized as indirect test because it is performed as no-load test and the losses and efficiency are calculated for different loading conditions. This test can not be performed on DC series machines. The circuit arrangement is shown in Fig.5.39.

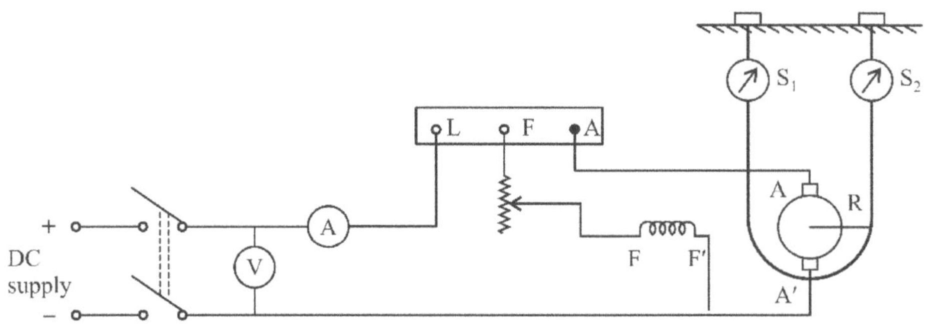

Figure 5.39 Swinburne Test

The machine is run at constant rated speed at rated voltage under no load condition. The losses that occur are as follows.

(i) Mechanical loss

(ii) Iron loss

(iii) $I_f^2 R_f$ loss

(iv) $I_{ao}^2 R_a$ loss

where I_f is the constant field current and R_f is the field winding resistance.

I_a is the no load armature current and R_a is the armature winding resistance.

The first three losses being, more or less constant in nature, together can be represented as P_{const}.

The input power at no load can be expressed as:

$$V_t (I_{ao} + I_f) = P_{costant} + I_{ao}^2 R_a$$

$$\therefore \quad P_{constant} = V_t (I_{ao} + I_f) - I_{ao}^2 R_a$$

With this information, now the efficiency can be calculated at any given load,

$$\text{Total loss} = P_{constant} + P_{variable}$$

where, $$P_{variable} = I_a^2 R_a$$

$$\therefore \quad \eta = 1 - \frac{P_{variable} + P_{constant}}{V_t I_L}$$

where $$I_L = I_a + I_f$$

The Swinburne's test is performed at no load hence the losses occurring during test are low. Further, the efficiency of the machine can be predetermined for any load. However, the effects that appear when the machine is actually loaded are not taken care of; such as brush contact loss, effect of temperature rise, variation in iron loss at pole – shoes etc.

5.12.3 Regenerative / Hopkinson's Test (Indirect Method)

Hopkinson's test is a full load, regenerative test performed by connecting two identical DC shunt machines, electrically back to back, and mechanically coupled with each other. The connection diagram is shown in Fig.5.40.

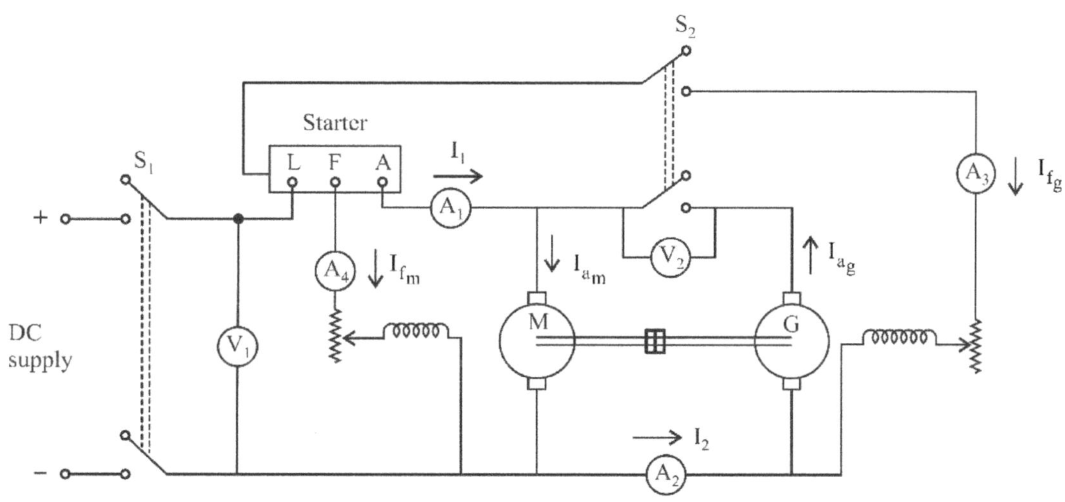

Figure 5.40 Hopkinson's Test

One of the machines is chosen to operate as a motor and is connected to a starter. The other machine would act as a generator, to which the first machine (motor) acts as the prime mover. The armature and field circuits of the machines are such that the generator feeds the electrical power to the motor while the motor feeds the mechanical power to the generator through this back to back arrangement. In a loss less system, no external power would be needed to run this system. Keeping the switch S_2 open, the dc supply is fed to the motor. Its field is adjusted to make it run at rated speed. This makes the generator also to run at rated speed as the two machines are identical. Switch S_2 is then closed and the field of the generator is adjusted such that the voltmeter V_2 reads nearly zero value. By doing this it is ensured that the terminal voltages across the two machines become equal.

A current I_2 starts circulating such that

$$I_{am} = I_2 = I_{ag}$$

It is important to note that the polarities of the two machines are rightly connected. The field of the motor is then slightly weakened so that current I_2 becomes equal to the rated current of the generator.

Under this condition the motor runs at its full load condition and gets the supply from the generator. The current I_1 flows to feed the losses of both the machines (say W_o).

The total loss $W_o = V_1 I_1$

Let stray loss of each machine be W_{stray}, then

$$2\,W_{stray} = W_o = (I_{am}^2 R_{am} + I_{ag}^2 R_{ag});\ \text{where } I_{am} = (I_1 + I_2)$$

Where the suffix used carry the usual meaning. The value of W_{stray} can thus be taken as stray loss for each machine.

$$\text{Efficiency of Motor } (\eta_m) = 1 - \frac{\text{loss}}{\text{input}}$$

where input $= V_1\, I_{am}$ and loss $= (W_{stray} + I_{am}^2 R_{am})$

Also, $I_{am} = (I_1 + I_2)$

Similarly,

$$\text{Efficiency of Generator } (\eta_g) = \frac{\text{output}}{\text{output+loss}}$$

where output $= V_1 I_{ag}$ and loss $= (W_{stray} + I_{ag}^2 R_{ag})$

Also, $I_{ag} = I_2$

This is an approximate method of finding out the losses and the efficiency as both the machines may not be exactly identical. The armature current of motor is higher, leading to higher armature reaction resulting in higher iron loss at its pole shoes.

Both the machines work on no load condition. The power taken from supply is very small, making this test economical. The temperature-rise and commutation problems can be checked by running the motor at rated load condition. This method is specially chosen when the sizes of the machines are large.

5.13 Concept of Braking in DC Motors

Any running DC Motor can be stopped by applying mechanical brakes. This causes a lot of friction, heat and loss of power. Electrical braking systems are, therefore, used to stop a DC motor at an exact position. It can also be used to control the speed during deceleration.

Three methods are available for electrical braking of DC motors.

(i) Rheostatic (Dynamic) braking,

(ii) Regenerative braking, and

(iii) Plugging

The concept of braking can be understood by Speed Torque Quadrants as shown in Fig.5.41.

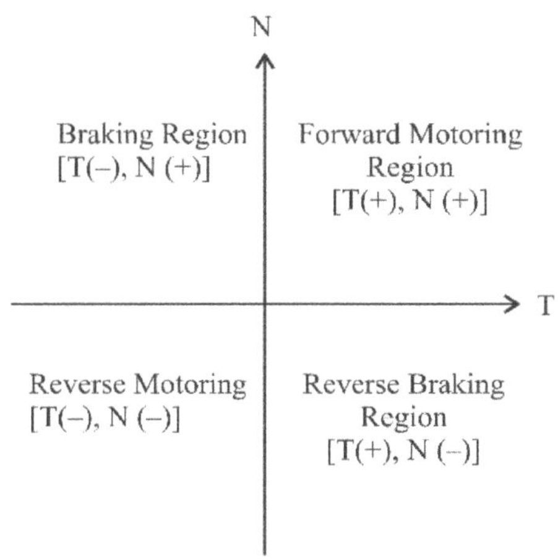

Figure 5.41 The four quadrants of operation of a motor

The concept of braking can further be elaborated by reconsidering the four–quadrant operation of Fig.5.41 for a separately excited DC machine as shown in Fig.5.42.

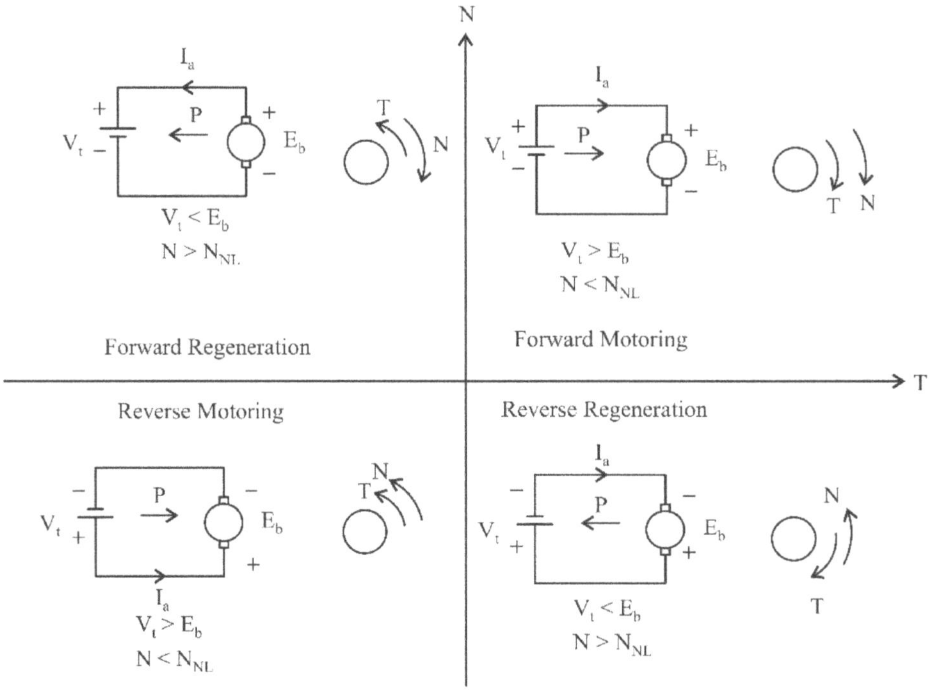

Figure 5.42 Four quadrant operation of separately excited DC motor

It is observed from Fig.5.42 that the direction of motor gets reversed by reversing the supply voltage. Also, the regenerative braking can be obtained in both the forward and reverse directions of rotation of the motor.

5.13.1 Rheostatic (Dynamic) Braking

The armature of the motor is disconnected from the supply and is short circuited through a pure resistance. The motor stops drawing power from the source. It keeps rotating due to stored kinetic energy and finally get stopped. In this duration the motor behaves as a generator. No additional power is required to stop the motor.

The circuit for rheostatic braking is shown in Fig.5.43.

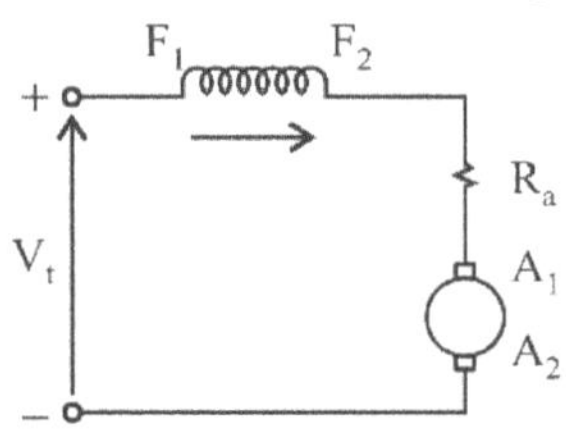

(a) Normal operation of DC series motor

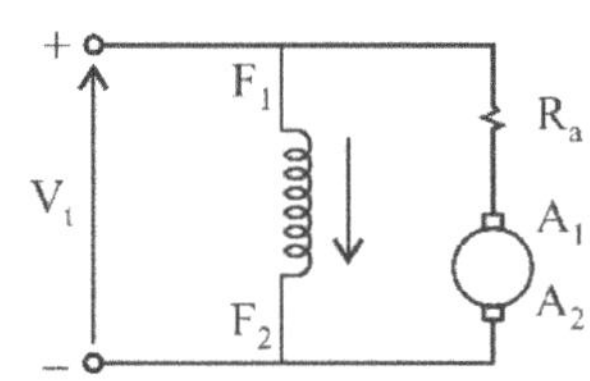

(b) Normal operation of DC shunt motor

(c) Braking of DC series motor (d) Braking of DC shunt motor

Figure 5.43 Rheostatic Braking of DC Motors

It is important to note that the direction of field current is kept the same. In this method the power gets wasted in the external resistance in the form of heat.

During rheostatic braking

$$E_b + (R_a + R)\, I_a = 0$$

Or $$I_a = \frac{-E_b}{(R_a + R)}$$

We know that

$$T \propto K_t\, \phi\, I_a$$

or $$T \propto K_t\, \phi\, \frac{E_b}{(R_a + R)}$$

or $$T \propto K_t\, \phi \left[\frac{K_n \phi N}{(R_a + R)}\right]; \text{ where } E_b \propto K_n\, \phi\, N$$

∴ $$T \propto \frac{K_t K_n}{(R_a + R)}\, \phi^2\, N$$

The speed torque characteristics of DC motors during braking are shown in Fig.5.44.

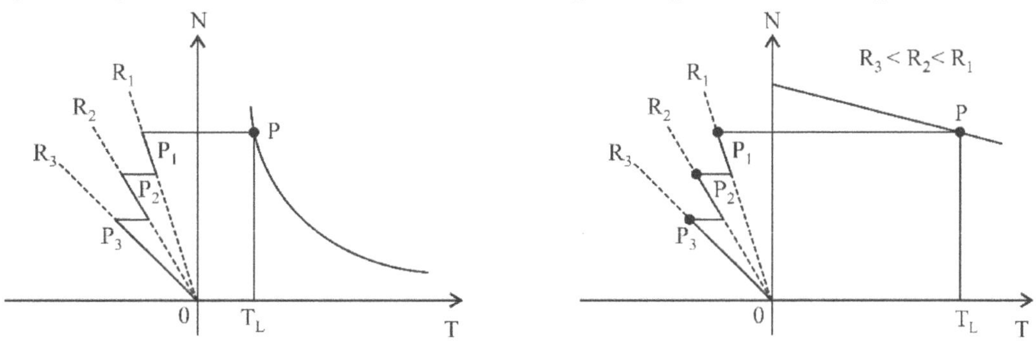

(a) DC series motor (b) DC shunt motor

Figure 5.44 Speed torque characteristic during Rheostatic Braking

At load torque T_L; the point of operation is 'P'. It gets shifted to P_1, P_2, P_3 to 0 while the rheostatic resistance is decreased from R_1 to R_2 to R_3.

5.13.2 Regenerative Braking

Regenerative braking is normally used where the speed of the motor is required to be maintained constant when the speed of the motor rises above the rated speed. For example, when a train runs down a negative gradient route or when a hoist lowers a load. This method of braking is not used to stop the motor. This method can be applied on motors that are designed to operate at speeds greater than no load speeds. The circuit and the speed torque characteristic are shown in Fig.5.45.

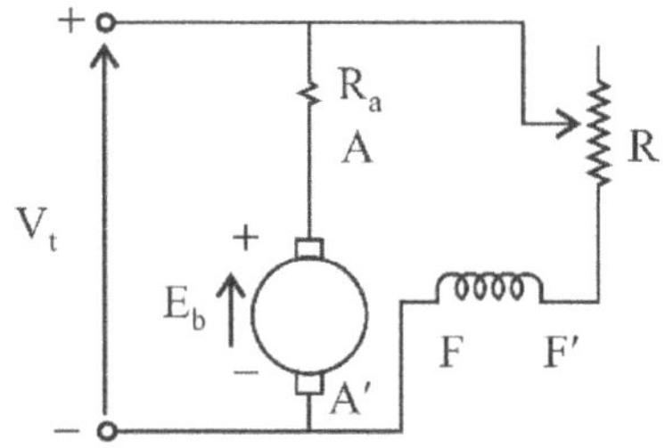

(a) Circuit arrangement

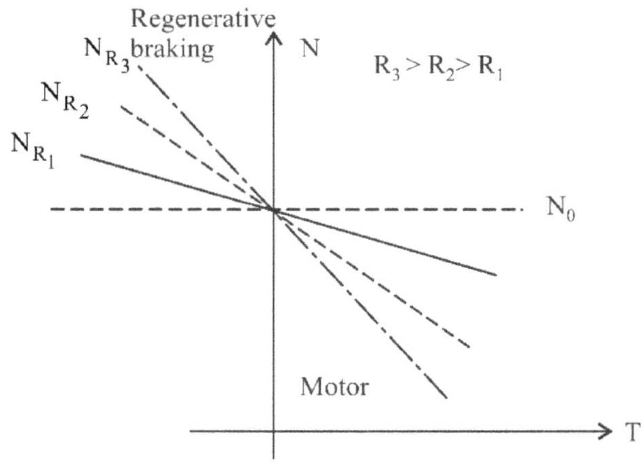

(b) Speed torque characteristics

Figure 5.45 Regenerative braking

In Regenerative braking the motor current gets reversed and hence the machine feeds the power back to the supply. It can not be applied to DC series motors as such.

5.13.3 Plugging (Reverse Current Braking)

The direction of armature current is reversed in order to bring the motor in braking mode. A DPDT (Double Pole Double Throw) switch is used to reverse the supply across armature, while it is ensured that the direction of field current does not change. This method therefore can be applied to shunt as well as series DC motors. The circuit arrangements are shown in Fig.5.46.

In order to understand the effect of reversal of supply, let us consider the normal circuit arrangement as shown in Fig.5.47.

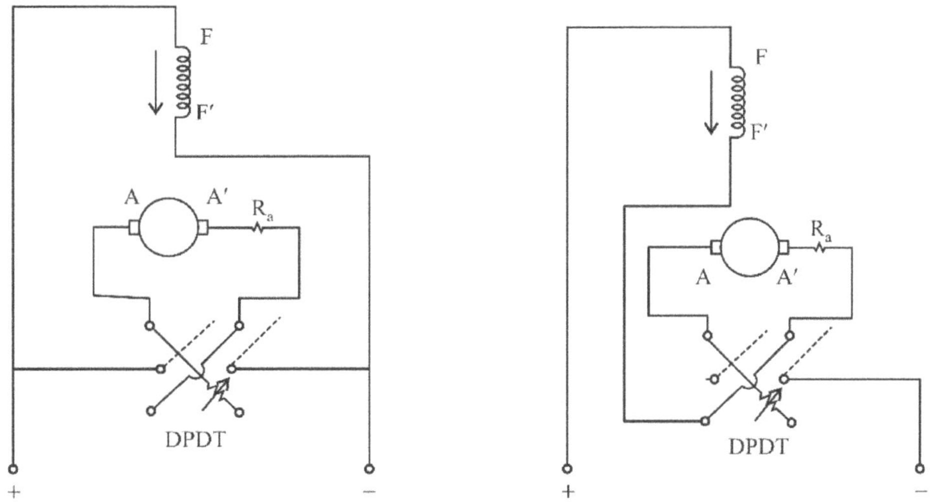

(a) DC shunt motor (b) DC series motor

Figure 5.46 Circuit for plugging

(a) Normal supply (b) Supply reversed

Figure 5.47 Immediate effect of supply reversal

The voltage current relationship can be given by

Under normal supply:

$$V_t = E_b + I_a R_a$$

Under Supply reversal:

The direction of E_b would not change immediately due to KE and due to the load. It can be said that at the time of reversal, $E_b \approx V_t$.

Hence in $V_t = -E_b + I_a R$, substitute for E_b as V_t

$$2V_t \approx I_a R_a$$

or $$I_a \approx \frac{2V_t}{Ra}$$

It means that nearly double the rated current will flow through the motor on supply reversal. Therefore, an external braking resistance R_B (nearly double the value of starting resistance) is inserted during supply reversal so that the current I_a is limited within limit, to a value I_B (braking current).

$$I_B \approx \frac{2V_t}{R_a + R_B}$$

The braking torque, $T_B \propto \phi I_B$, will start acting on motor. The speed of motor starts decreasing to a value equal to zero.

The supply must be disconnected as soon as the motor comes near to its stand still position. Otherwise the motor starts running in opposite direction (reverse motoring mode).

5.14 Applications of DC Motors

A summary of characteristics and applications of different types of DC motors is given in the following table.

Table 5.1

Type of Motor	Characteristics	Applications
Series	(a) Very high starting torque (b) Varying speed, being dangerously high at no load (c) Self-adjustable varying speed at load.	(i) For traction work i.e., electric loco motive, trolley cars and buses. (ii) Cranes and hoists (iii) Conveyors (belt)
Shunt	(a) Nearly constant speed (b) Medium starting torque (1.5 to 2 times full load torque	(i) For driving constant speed shafts. (ii) Lathes, centrifugal pumps, reciprocating pumps. (iii) Fans and blowers (iv) Wood working machines, paper mills and printing presses.
Cumulative compound	(a) High starting torque (b) Varying speed but within limit (c) Self-adjustable varying speed at load	(i) For intermittent high torque load (ii) Shears, punches and presses (iii) Elevators conveyors and stokers (iv) Heavy rolling mills
Differential compound	(a) Constant speed but danger of running away (b) Low starting torque	Not generally preferred. Sometimes a very few turns of differentially connected series field is used to maintain constant speed.

REVIEW QUESTIONS

1. Develop the expression of electromagnetic torque in a DC machine.
2. Deduce the expression of speed of a DC motor and hence suggest the methods of speed control.
3. Draw the speed – torque characteristics of DC series motors, DC shunt motors and the compound motors.
4. Explain Ward-Leonard method of speed control of DC motors.

5. Explain the need of starter for DC motors.
6. What are the functions performed by a three-point starter?
7. How does a four-point starter differ from a three-point starter? Give its merits and demerits.
8. How can the efficiency of DC motors be found without actually loading them?
9. State purpose of Electric Braking and name various types of Electric Braking.
10. Explain Plugging with suitable diagrams for a DC motor. What care should be taken while applying Plugging?
11. Discuss Dynamic Braking for various types of motors.
12. Discuss about the applications of DC motors.

PROBLEMS

1. The armature of a 240 V DC shunt motor, running at 800 rpm, draws a current of 2 A. By field control method, it is made to run at 950 rpm while drawing line current of 30 A. How much field resistance is required to be put in series, if the values of field winding and armature winding resistances are160 ohm and 0.4 ohm respectively? [38.2 ohm].
2. A 230 V DC shunt motor draws armature current of 20 A at some load. The armature resistance is 0.5 ohm. How much series resistance is required to be inserted in armature circuit to reduce the speed by 50% if, (a) the load torque is constant and (b) the load torque is proportional to the square of the speed? [5.5 ohm, 23.5 ohm]
3. A DC shunt machine generates 250 volts on open circuit at 1000 rpm. The armature resistance is 0.5 ohm and field resistance is 250 ohm. The machine runs as a motor at no load and draws a current of 4 A at 250 V. What will be the efficiency and speed of the motor when it draws 40 A at 250 V? The armature reaction weakens the flux by 4%. [960 rpm, 82.5%].
4. A 200 V shunt motor develops 23 hp when taking 20.2 kW. The armature and field resistances are 0.06 ohm and 50 ohms respectively. Find the efficiency and the power input when the output is 10 hp.

   [74.6%, 10 kW]
5. A 220 V shunt motor has an armature resistance of 0.5 ohm and takes a current of 40 A on full load. By how much must the main flux be reduced to raise the speed by 50%, if the developed torque is constant? [87.5%]
6. A 240 V series motor takes 40 A when giving its rated output at 1500 rpm. Its resistance is 0.8 ohm. What resistance must be added to obtain rated torque (a) at starting, (b) at 1000 rpm? [(a) 5.7 ohm, (b) 1.9 ohm]
7. A 500 V series motor with a resistance of 0.5 ohm runs at 160 rpm for I = 40 A, and at 140 rpm for I = 50 A. Calculate (a) total resistance R at starting for I = 50 A, (b) speed with R in circuit when I = 40 A. (c) total resistance with speed as in (b) and I = 50 A

   [(a) 10 ohm, (b) 83.8 rpm (c) 7.74 ohm]

8. A 240 V motor on brake test takes 52 A when running at 1500 rpm. The spring balance, at the end of 76 cm brake arm, reads 93.5 N. Find out the efficiency of the motor.

[90%]

9. A 480 V, 25 hp, shunt motor takes 2.5 A when running at no load. The armature and field resistances are 0.6 ohm and 800 ohms respectively. Brush drop is 2 V. Find the efficiency at full load.

[88.5%]

10. The Hopkinson's test on two DC shunt machines, each having armature resistance of 0.02 ohm; running at full load, gave the following results. Line voltage, 250 V; Line current (excluding field current), 50 A; motor armature current, 880 A; Field currents, 5A and 4.2 A. What is the efficiency of each machine?

[Motor 92.04%, Generator 92.02%]

www.ingramcontent.com/pod-product-compliance
Ingram Content Group UK Ltd.
Pitfield, Milton Keynes, MK11 3LW, UK
UKHW060019200726
13845UKWH00046B/91

9 789395 038263